BICENTENNIAL
1807
WILEY
2007
BICENTENNIAL

THE WILEY BICENTENNIAL—KNOWLEDGE FOR GENERATIONS

$\mathcal{E}$ach generation has its unique needs and aspirations. When Charles Wiley first opened his small printing shop in lower Manhattan in 1807, it was a generation of boundless potential searching for an identity. And we were there, helping to define a new American literary tradition. Over half a century later, in the midst of the Second Industrial Revolution, it was a generation focused on building the future. Once again, we were there, supplying the critical scientific, technical, and engineering knowledge that helped frame the world. Throughout the 20th Century, and into the new millennium, nations began to reach out beyond their own borders and a new international community was born. Wiley was there, expanding its operations around the world to enable a global exchange of ideas, opinions, and know-how.

For 200 years, Wiley has been an integral part of each generation's journey, enabling the flow of information and understanding necessary to meet their needs and fulfill their aspirations. Today, bold new technologies are changing the way we live and learn. Wiley will be there, providing you the must-have knowledge you need to imagine new worlds, new possibilities, and new opportunities.

Generations come and go, but you can always count on Wiley to provide you the knowledge you need, when and where you need it!

WILLIAM J. PESCE
PRESIDENT AND CHIEF EXECUTIVE OFFICER

PETER BOOTH WILEY
CHAIRMAN OF THE BOARD

Study Guide

Rachael Henriques Porter
Le Moyne College

Foundations of College Chemistry

12th Edition
Alternate 12th Edition

Morris Hein
Mount San Antonio College

Susan Arena
University of Illinois
Urbana–Champaign

John Wiley & Sons, Inc.

Cover Photo: Digital Vision/Getty Images

To order books or for customer service please, call 1(800)-CALL-WILEY (225-5945).

ISBN-13 978-0-470-06716-1

ISBN-10 0-470-06716-0

Printed in the United States of America

10 9 8 7 6 5 4 3 2 1

Printed and bound by Malloy Lithographing, Inc.

For my boys, K, K, C, C

–R.H.P.

Preface

This study guide has been prepared to accompany *Foundations of College Chemistry, Twelfth Edition* and *Foundations of College Chemistry, Alternate Twelfth Edition*, both by Hein and Arena. The guide selects certain key concepts from the text, and provides the student with a means of self-evaluation for determining how well he or she understands the material of each chapter.

By presenting slightly different viewpoint and emphasis, the study guide provides students with an approach other than that of the textbook toward mastery of the subject matter. Because this guide is an auxiliary student-oriented aid to complement the textbook, no new material is presented.

The chapters in the guide are organized as follows: The chapter headings are the same as in the text and are followed by a section called "Selected Concepts Revisited" that recaps some of concepts that many students in the past have had difficulty grasping. A "Common Pitfalls to Avoid" section is intended to make students aware of common mistakes previously made by students. By reading this section before completing the self-evaluation section hopefully students will avoid making similar mistakes. The self-evaluation section provides various exercises that allow the student to check his or her understanding of the text chapter material. Students are asked to complete fill-in responses or to choose an appropriate response. Tables to be constructed and problems to be solved are presented. Nomenclature, equation balancing, and use of the periodic table are covered. The last problems in the self-evaluation section are identified as challenge problems. These will usually be a little more difficult or complex than the other exercises in the chapter. Additionally, the explanations about the answers will be briefer and will assume that the student has a good grasp of the basic material covered up to that particular point. Complete answers to questions and solutions to problems are given at the end of each chapter, to provide immediate reinforcement or correction of errors. The last section in each chapter provides a recap or summary.

In addition to the usual exercises, several puzzles are included in the guide to give the student a break and to check his or her vocabulary skills. The student will probably need a periodic table to help in solving the puzzles.

The authors visualize using the study guide as follows. The student would (1) read the assigned text chapter; (2) go through the associated chapter in the study guide including the self-evaluation process; (3) return to the assigned homework problems, lab experiments, and other instructor-generated activities; and (4) be tested on predetermined performance objectives for each chapter.

To the Student

This guide, which accompanies the textbook *Foundations of College Chemistry, Twelfth Edition* and *Foundations of College Chemistry, Alternate Twelfth Edition*, both by Hein and Arena, is a student-oriented self-study guide. It has been prepared to help you evaluate your grasp of certain concepts in the textbook; you can then identify trouble spots and ask your instructor for individual help.

What is the place of the study guide in the total chemistry program? Your instructor will hand out reading assignments from the textbook, which will be followed by homework questions and problems and laboratory experiments. After classroom discussion and problem sessions, you will be tested on the material. So, when do you use the study guide? A typical student might first read the assigned chapter in the text (with rereading, if necessary), and then turn to the study guide to check for common misconceptions or easily avoidable mistakes. Completing the self-evaluation section in the study guide then serves as a check for your understanding of the key concepts. All of the answers and solutions are given so that any errors can be quickly spotted. If a particular concept or type of problem doesn't make sense, the student can return to the text-book. Then the student is ready to work the homework assignment, knowing that he or she understands the basic concepts in the chapter. The week's lab experiment is performed, followed by the chapter quiz.

Each chapter of the study guide has questions to answer and problems to solve and is organized as follows: The "Selected Concepts Revisited" section provides students with the opportunity to review some (not all) of the concepts presented in the chapter that have traditionally confused students or are more emphasized, and provides some helpful hints for dealing with the material. The next section "Common Pitfalls to Avoid" highlights some of the mistakes students typically make in answering questions related to this chapter's topics. By making you aware of these possible errors, it is hoped that you will be less likely to repeat these mistakes. The "Self-Evaluation Section" provides various types of exercises that enable you to check your understanding of the key objectives. The answers and solutions to the problems are given at the end of the chapter so that you can find out where your difficulty may have occurred. The challenge problems at the end of each chapter are usually a bit more difficult or complex than the rest of the self-evaluation exercises. You should work on these after you feel comfortable with the other material. The answers to the challenge problems may not be as thorough as the other answers since you're expected to have a good grasp of the basic material before attempting the challenge problems. The last section gives a brief recap of the chapter and indicates where you are going next in the course.

The study guide allows you to work on your own and make efficient use of your time. It should assist you in finding out whether you have learned the basic concepts.

Best wishes for success in your study of chemistry.

Contents

1 An Introduction to Chemistry. 1

2 Standards for Measurement 7

3 Elements and Compounds 19

4 Properties of Matter 25

5 Early Atomic Theory and Structure 33

6 Nomenclature of Inorganic Compounds 39

7 Quantitative Composition of Compounds 45

8 Chemical Equations 59

9 Calculations from Chemical Equations 69

10 Modern Atomic Theory and the Periodic Table. 79

11 Chemical Bonds: The Formation of Compounds from Atoms 87

12 The Gaseous State of Matter 99

13 Properties of Liquids 113

14 Solutions. 123

15 Acids, Bases, and Salts. 145

16 Chemical Equilibrium 155

17 Oxidation–Reduction 175

18 Nuclear Chemistry 189

19 Introduction to Organic Chemistry. 195

20 Introduction to Biochemistry 213

CHAPTER ONE

An Introduction to Chemistry

SELF-EVALUATION SECTION

We will begin your use of the study guide with some straight-forward exercises followed by a quick mathematics section. As mentioned in the introduction, the use of mathematics is very important for your success in chemistry.

1.   The scientific method is a general approach to answer questions or solve problems. Each of the following statements is missing an "action verb" which is to be selected from the list below.

Use these verbs to fill in the blanks:

        formulate
        modify
        analyze
        plan and do
        collect

(1)   _____ the facts or data that are relevant to the question which is usually done by experimentation.

(2)   _____ the data to find trends.

(3)   _____ a hypothesis that will account for the accumulated data and can be tested by further experimentation.

(4)   _____ additional experiments to test the hypothesis.

(5)   _____ the hypothesis as necessary so that it is compatible with all pertinent data.

2. Match the correct definition to each of the following terms.

    (1)   macroscopic   _____   molecular view of substances

    (2)   amorphous   _____   indefinite shape, definite volume

    (3)   mixture   _____   the big picture; everyday objects

    (4)   gas   _____   containing two or more physically distinct phases

    (5)   microscopic   _____   a solid lacking a regular internal geometric pattern

    (6)   liquid   _____   indefinite shape, no fixed volume

    (7)   heterogeneous   _____   contains two or more substances

3. Fill in the blank.

    (1)   If a solid is not amorphous, it is _____.

    (2)   If a mixture is not heterogeneous, it is _____.

    (3)   If a material is not a mixture, it is a _____ _____.

4. All matter in the universe can be classified into one of three states – gas, liquid, or solid. Determine to which of the three states of matter each of the following descriptions relates.

Description	States of Matter
(1) Has a definite fixed shape.	_____
(2) Particles flow over each other while retaining fixed volume.	_____
(3) Exerts a pressure on all walls of the container.	_____
(4) Exhibits no or very slight compressibility.	_____
(5) Particles move independently of each other.	_____
(6) Particles arranged in regular, fixed geometric pattern.	_____
(7) Exhibits slight compressibility.	_____
(8) Exhibits very high compressibility.	_____

5. Examine the following list of items. Some are mixtures, others are elements or compounds. Place an *S* next to the pure substance and an *M* next to the mixtures.

(1) wood _____ (6) water _____

(2) sodium chloride _____ (7) air _____

(3) milk _____ (8) chlorine _____

(4) oxygen _____ (9) soil _____

(5) rubber _____ (10) gasoline _____

6. As a good way to check your readiness to begin a course in chemistry, work through the following mathematical problems. Many students have found it helpful in determining areas to be reviewed.

Circle the response for each of the questions.

(a) The problem is worked correctly.
(b) The problem has been worked incorrectly.

**Do your calculations
in this space
or on scratch paper**

(1) $2/4 = 0.33$ a   b

(2) $1/16 = 0.0625$ a   b

(3) $1/10 = 0.01$ a   b

(4) $30/6 = 0.5$ a   b

(5) $1/3 \times 4/5 = 5/8$ a   b

(6) $1/2 \div 1/4 = 2$ a   b

(7) $3/4 + 3/5 = 3/20$ a   b

(8) $1/3 - 1/4 = 1/12$ a   b

(9) $\dfrac{2 \times 4 \times 3}{6 \times 2} = 2$ a   b

(10) $10^2 = 100$ a   b

(11) $\sqrt{25} = 5$ a   b

(12) $2^3 = 6$ a   b

(13) $\sqrt[3]{8} = 2$ a   b

(14) $6^2 \times 6^3 = 6^6$ a   b

(15) $\dfrac{9^4}{9^3} = 9$ a   b

(16) $\dfrac{a^7}{a^{10}} = a^{-3}$ a   b

(17) $6.9 \times 10^3 = 690$ a   b

(18) $0.0054 = 5.4 \times 10^{-3}$ a   b

(19) $(2 \times 10^3)(5 \times 10^2) = 10^6$ a   b

(20) $(5 \times 10^{-3})(3 \times 10^5) = 1.5 \times 10^2$ a   b

(21) 1594.61 rounded off to four figures is 1596   a   b

Do your calculations
in this space
or on scratch paper

(22) The reciprocal of 2 is 1/2   a   b

(23)   6.1 cm
       $\times$ 3.5 cm
        305
        ───
        183

~~21.35 cm²~~ Is this answer written
with the correct number of
significant figures?   a   b

(24) If $\dfrac{x}{2.5} = \dfrac{3}{7.5}$, then $x = 1$ a   b

(25) Let $T_1 \times N_1 = T_2 \times N_2$
If $T_1 = 5$, $N_1 = 3$ and $T_2 = 6$,
then $N_2 = 2.5$ a   b

(26) If $\dfrac{P_1 V_1}{T_1} = \dfrac{P_2 V_2}{T_2}$, then $T_2 = \dfrac{P_2 V_2}{P_1 V_1 T_1}$ a   b

(27) If $\dfrac{1.5}{x} = \dfrac{3.0}{6.4}$, then $x = 3.5$ a   b

(28) If it takes 1 calorie to heat 1 gram
of water 1°C, how many calories
will be required to heat 3 grams of
water 5°C? Answer: 8 calories   a   b

Check your answers. Students who miss more than half of the problems will likely have a difficult time with the mathematically related material in the text without an intensive review and extra work outside of class. Keep this in mind if you found that many of the problems were unfamiliar to you or you responded incorrectly. If you missed more than a quarter of the problems, you should study thoroughly the Mathematical Review Appendix at the back of the text.

RECAP SECTION

The objectives of Chapter 1 are to introduce you to the broad field of science known as chemistry. We discuss the importance of chemistry in other fields of science such as biology and agriculture.

You should also gain some insight into how the scientific method is applied in chemistry. As you study chemistry or any experimental science, you find the steps of hypothesis, theory, and scientific law leading time and again toward an understanding of natural phenomena.

The importance of problem solving to the science of chemistry cannot be over-emphasized. You will be working problems within the text material, solving homework problems, and using this study guide to evaluate your ability to solve certain specific types of problems. Many of the laboratory experiments also involve problem solving.

The techniques you will use are mostly simple arithmetic operations, but you must have a complete understanding of them. Significant figures and scientific notation are also important for your progress in chemistry. Make sure you know how to work your calculator too!

The topics of scientific notation, significant figures and rounding-off numbers are covered thoroughly in Chapter 2 of the text. Remember that additional helpful information which will be of use throughout your study of chemistry is found in Appendix I of the text.

ANSWERS TO QUESTIONS

1. (1) collect   (2) analyze   (3) formulate   (4) plan and do   (5) modify

2. (5), (6), (1), (7), (2), (4), (3)

3. (1) crystalline   (2) homogeneous   (3) pure substance

4. (1) solid   (2) liquid   (3) gas   (4) solid
    (5) gas   (6) solid   (7) liquid   (8) gas

5. (1) M   (2) S   (3) M   (4) S
    (5) M   (6) S   (7) M   (8) S
    (9) M   (10) M

6. (1) b   (8) a   (15) a   (22) a
    (2) a   (9) a   (16) a   (23) b
    (3) b   (10) a   (17) b   (24) a
    (4) b   (11) a   (18) a   (25) a
    (5) b   (12) b   (19) a   (26) b
    (6) a   (13) a   (20) b   (27) b
    (7) b   (14) b   (21) a   (28) b

CHAPTER TWO

Standards for Measurement

SELECTED CONCEPTS REVISITED

Numbers expressed in scientific notation essentially have two parts:
- the part showing the significant figures (generally only one digit before the decimal)
- the part showing the location of the decimal point (the relative magnitude of the number). A negative power of 10 means the original number is less than 1.

An important issue involved with all these measurements is significant figures. Bear in mind the information found in significant figures. Significant figures essentially indicate the precision to which an instrument was read or a calculation was based on. The last digit of a measurement has some degree of uncertainty and, in direct measurements, is often estimated. Exact numbers have no degree of uncertainty (that is, they are known exactly) and therefore have infinite significant figures.

Most often significant figures will be limited by numbers read in a measurement. Zeros in significant figures can be summarized by looking at these two examples; numbers in bold are significant.

0.00**45060**   **301**000

Leading zeros are NOT significant; middle zeros are ALWAYS significant; *trailing zeros are significant ONLY when a decimal point is present.*

Counted integers and defined numbers are exact numbers (no uncertainty).

Quantitative is related to a quantity; a measurement or an amount.

Qualitative is related to description.

You will find it invaluable to carry out some actual measurements to better visualize the relative sizes of metric versus standard units. Some relationships that you may find helpful include the following. The 1500 m race is often called the metric mile (note that 1609 m is actually a more precise conversion). If you buy gas for your car in Canada you can multiply the listed price per liter by four to find the approximate price per gallon (a more precise conversion would involve multiplying by 3.78).

Two terms introduced in Chapter 2 that are used in everyday conversation are *heat* and *temperature*. There is some confusion about the proper usage of *temperature* and *heat*. It is usual to state that the temperature of an object indicates how much heat the object contains. But this is an incorrect notion. Think of two objects of different size but at the same temperature. Which object contains more heat energy? The larger object, of course. Thus, temperature is only an indication of the degree of "hotness" or "coldness" of an object. The temperature scales are relative, arbitrary scales established with certain reference points. The Celsius scale, for instance, has 0 degrees set at the freezing point of water and 100 degrees set at the boiling point of water.

When converting between units of temperature, consider whether you need to correct only for the difference in the zero point on the temperature scale or if the size of the degree also needs correcting. Converting between Celsius and Kelvin requires a zero point correction; a simple addition or subtraction of the difference in the zero point (273.15). When converting between Fahrenheit and Celsius, both zero point and degree size corrections are needed. An addition or subtraction term (of 32) takes care of the zero point correction, and a multiplication or division term (by 1.8) takes care of the size of the degree correction.

The last term defined in this chapter is *density*. Again, we can use either an algebraic form of definition or the English equivalent. The density of a substance is related to its mass and volume. Since all matter in the universe has mass and occupies space or volume, the quantity defined as density is a useful physical characteristic used in describing various substances. The defining equation for density says that the density of a substance is equal to the mass divided by the volume. The mass is given in grams, and the volume can be expressed in cubic centimeters, millimeters, or liters. The units for density are therefore either grams/cubic centimeter (g/cm^3), grams/milliliter (g/mL), or grams/liter (g/L).

COMMON PITFALLS TO AVOID

Do not forget to check the logic of your answer, especially with unit conversion problems. If you measure your height in centimeters and after converting to feet, get 18 ft as your answer, you did something wrong. Look to see if your answer makes sense; should it be smaller or larger than the original number? If one pound is less than one kilogram then if you convert 145 lbs. to kg the number should be less than 145. One method of double-checking your answer is to take the answer you calculated and independently do the problem in reverse to see if you get the original number.

Many students learn the prefixes for the metric units but then get confused when actually using them in a conversion. For example, to convert from grams (g) to milligrams (mg), they know milli means 10^{-3} (or 0.001) but they forget whether $1 g = 10^{-3} mg$ or $1 mg = 10^{-3} g$. Again, think it through logically. If you know a mg is smaller than a gram, then "1" of the smaller unit must be equal to a really small portion of the larger unit (or it takes many small units to equal one larger unit). Therefore $1 mg = 10^{-3} g$, i.e., $1 mg = 0.001 g$, or $1000 mg = 1 g$.

Keep track of units in a calculation and whether they are numerators or denominators. You should be able to see clearly which units get cancelled and which are left at the end of the calculation. For example, please remember that $g/(g/cm^3)$ will give you units of cm^3.

Be careful when rounding off numbers, e.g. rounding 1846 to two significant figures (it is not 18). Think of this number as though it were money. If you were owed $1846, would you want it rounded off to $18 or $1800? The trailing zeroes are necessary to show placement of the decimal point while still showing those digits are not significant.

When carrying out a calculation stepwise be careful if you have a habit of rounding off at each step. (Check with your instructor to find out their policy.) In general it is better and more precise to carry a couple extra significant figures to the end of the calculation and save the full rounding off to the end of the calculation.

1. In order to handle large numbers efficiently, it is worthwhile to use scientific notation, which means writing a number as a power of 10. The numbers are always written between 1 and 10 with the associated power of 10.

    (1) Express the following numbers as powers of 10:
        0.0305   _____
        29   _____
        0.000721   _____
        680,000,000   _____
        36,800   _____

    (2) Express the following in decimal form:
        4.77×10^4 _____
        8.41×10^{-2} _____
        5.8×10^1 _____
        9.1×10^0 _____
        1.415×10^2 _____

2. Express the following numbers in exponential form:

    (1) 67,000 _____   (4) 0.00078 _____

    (2) 0.0654 _____   (5) 411,000 _____

    (3) 10,000,000 _____   (6) 10 _____

3. Express the following numbers in decimal form:

    (1) 4.8×10^3 _____   (4) 1074×10^{-2} _____

    (2) 0.67×10^2 _____   (5) 0.0034×10^3 _____

    (3) 0.151×10^{-5} _____   (6) 11×10^{-5} _____

4. Round off the following numbers to four significant digits:

    (1) 32.983 _____   (5) 0.182710 _____

    (2) 101.049 _____   (6) 69.93606 _____

    (3) 2.086699 _____   (7) 0.0026155 _____

    (4) 4.0000095 _____   (8) 83.99998 _____

5. Three quantities routinely measured include (1) _____, (2) _____, and (3) _____. These basic measurements allow us to calculate many other quantities such as density which is (4) _____/ (5) _____.

The kilogram, abbreviated (6) _____, is used to measure (7) _____.

(8) _____ is measured in (9) _____, abbreviated m.

However, probably the most common type of measurement we make in a chemistry laboratory involves the metric volume (10) _____, which is abbreviated (11) _____.

To these three units of measure for mass, length, and volume, we can attach prefixes to change the unit's value by various powers of 10. For example, *kilo* means (12) _____ and *milli* means (13) _____. The prefix for 0.01 (10^{-2}) is (14) _____, and the prefix for 0.000001 (10^{-6}) is (15) _____. The abbreviations for the pre-fixes *kilo, centi, micro*, and *milli* are (16) _____, (17) _____, (18) _____, and (19) _____.

Using the four prefixes in general use in chemistry, we must be able to convert from one metric unit to any other corresponding metric unit. We will use a series of conversions to find out if a series of decimal-point changes presents any problems to you.

Convert These Units   **To These Units**

(20) 42 g   _____mg

(21) 0.741 cm   _____m

(22) 8.4 mL   _____L

(23) 776 kg   _____g

(24) 1005 μL   _____mL

(25) 15 μg   _____mg

(26) 6.2 L   _____mL

(27) 0.34 km   _____cm

Convert These Units   **To These Units**

(28) 1.25 mg   _____g

(29) 0.0089 g   _____mg

(30) 9 cm   _____μm

6. Solve for x.

    (1) $x = (1.3 \times 10^2)(6.4 \times 10^{-1})$

    (2) $x = (5.76 \times 10^{-4})(2.4 \times 10^3)$

    (3) $(8.9 \times 10^1)\, x = (5.46 \times 10^3)$

    (4) $(0.77 \times 10^4)\, x = 125{,}000$

7. Converting American units to metric units.

    (1)   A college basketball player playing forward is generally about 6 ft 6 in. tall. How many centimeters is this? How many meters?

        Do your calculations here.

    (2)   Tall cans of juice contain 46.0 fluid ounces. How many mL is this? One ounce equals 29.6 mL.

        Do your calculations here.

    (3)   A can of green beans costs 59¢ and weighs 1 pound. What is the cost in cents per gram? The words "cents per gram" are a way of expressing the fraction cents/gram. This means that the number of cents is divided by the number of grams. The word "per" is a common method used in science to indicate a division operation or a fraction. Be alert to this term in future problems.

           1 lb = 453.6 g

        Do your calculations here.

8. We can convert the Kelvin temperature scale to the Celsius scale, and vice versa, with the formula K = °C + 273.15. The value 273.15 is a constant value. Using the formula, make the following conversions.

Convert From   **To**

(1)   273K   _____ °C

(2)   −15°C   _____ K

(3)   −35°C   _____ K

(4)   25°C   _____ K

(5)   348K   _____ °C

(6)   205°C   _____ K

(7)   29K   _____ °C

Using the equations given for converting °F to °C, and vice versa, make the conversions asked for.

$$°F = (1.8 \times °C) + 32$$

$$°C = \frac{(°F - 32)}{1.8}$$

(8)   The lowest temperature in the world for an inhabited area was -90.4°F in Siberia. The record high for this location is 94°F. How many degrees Celsius apart are these two temperatures?

Do your calculations here.

(9)   Surface temperatures on Venus are upwards of 450°C. What is that temperature in °F?

Do your calculations here.

9. As you progress in your study of various substances, you will find a great diversity in the physical nature of matter. There are many ways to categorize matter in trying to relate the properties of one substance to another. One such way is through a relationship called *density*. The density of a substance is not only related to its mass but also to the volume it occupies. Thus, a cube of aluminum metal measuring 10 cm on a side has much less mass than a cube of iron metal of the same volume. We say the density of aluminum is less than that of iron.

The equation defining density is written as follows:

$$\text{Density} = \frac{\text{mass}}{\text{volume}}$$

$$d = \frac{\text{g}}{\text{cm}^3} = \frac{\text{g}}{\text{mL}} \text{ or } \frac{\text{g}}{\text{liter}}$$

The units for density vary depending on the units used for volume.

(1) Sample problem: Our block of aluminum, 10 cm on a side, was found to have a mass of 2700 g by using a balance. What is the density of aluminum?

Do your calculations here.

(2) Another problem: Logs of black ironwood, which do not float in water, are found to have a density of 1.077 g/cm³. What is the volume of a piece of ironwood that has a mass of 750 g?

Do your calculations here.

Challenge Problems

10. The standard of metric length is the meter which was defined (prior to 1983) as 1,650,763.73 wavelengths of a spectral line of the element krypton. To three significant figures determine this wavelength in nanometers if $1 \text{ nm} = 10^{-9}$ m.

Do your calculations here.

11. For each of the following, identify (a) what is the numerical value and what is the unit, (b) how many significant figures, (c) which numbers are known and which are estimated. Round each number to two significant figures (d), and (e), express that number as a power of 10.

   (1)   1055 kg

   (2)   1,650,763 wavelengths

   (3)   1.077 g/cm^3

   (4)   0.00845 L

   (5)   776,000 g

RECAP SECTION

This has been a long chapter, but if you feel comfortable with the new terms and techniques, you've come a long way toward being successful in the remaining weeks of the course. We have reviewed some basic terms and concepts, brushed up on the metric system, and practiced converting one metric unit to another. Common methods and units of measuring mass, volume, and length were covered, as was density and its use in differentiating between substances. The temperature scales of the most concern are the Celsius and Kelvin scales. Significant figures, units of measurement, and scientific notation will be used time and again throughout this course. Mastering this material, utilizing your algebraic skills, is crucial to handling chemical problems in future chapters. If you had difficulty with this chapter, you should take the time now to go back over the Math Review in the text or seek extra help, before returning to the study guide.

ANSWERS TO QUESTIONS AND SOLUTIONS TO PROBLEMS

1. (1) 3.05×10^{-2}; 2.9×10^1; 7.21×10^{-4}; 6.8×10^8; 3.68×10^4
    (2) 47,700; 0.0841; 58; 9.1; 141.5

2. (1) 6.7×10^4 (4) 7.8×10^{-4}
    (2) 6.54×10^{-2} (5) 4.11×10^5
    (3) 1×10^7 (6) 1×10^1

3. (1) 4800   (4) 10.74
    (2) 67   (5) 3.4
    (3) 0.00000151   (6) 0.00011

4. (1) 32.98   (4) 4.000   (7) 0.002616
    (2) 101.0   (5) 0.1827   (8) 84.00
    (3) 2.087   (6) 69.94

5. (1), (2), (3) length, mass, volume (in any order)
    (4) mass, (5) volume, (6) kg, (7) mass, (8) length, (9) meters, (10) liter, (11) L
    (12) 1000   (13) 0.001   (14) centi-   (15) micro-   (16) k   (17) c   (18) μ (19) m

(20) 4.2×10^4 mg $\qquad$ $(42 \text{ g})\left(\dfrac{1000 \text{ mg}}{1 \text{ g}}\right) = 42,000 \text{ mg} = 4.2 \times 10^4 \text{ mg}$

(21) 0.00741 m $\qquad$ $(0.741 \text{ cm})\left(\dfrac{1 \text{ m}}{100 \text{ cm}}\right) = 0.00741 \text{ m}$

(22) 0.0084 L $\qquad$ $(8.4 \text{ mL})\left(\dfrac{1 \text{ L}}{1000 \text{ mL}}\right) = 0.0084 \text{ L}$

(23) 7.76×10^5 g $\qquad$ $(776 \text{ kg})\left(\dfrac{1000 \text{ g}}{1 \text{ kg}}\right) = 776,000 \text{ g} = 7.76 \times 10^5 \text{ g}$

(24) 1.005 mL $\qquad$ $(1005 \text{ } \mu\text{L})\left(\dfrac{1 \text{ mL}}{1000 \text{ } \mu\text{L}}\right) = 1.005 \text{ mL}$

(25) 0.015 mg $\qquad$ $(15 \text{ } \mu\text{g})\left(\dfrac{1 \text{ mg}}{1000 \text{ } \mu\text{g}}\right) = 0.015 \text{ mg}$

(26) 6.2×10^3 mL $\qquad$ $(6.2 \text{ liter})\left(\dfrac{1000 \text{ mL}}{1 \text{ liter}}\right) = 6200 \text{ mL} = 6.2 \times 10^3 \text{ mL}$

(27) 3.4×10^4 cm $\qquad$ $(0.34 \text{ km})\left(\dfrac{1000 \text{ m}}{1 \text{ km}}\right)\left(\dfrac{100 \text{ cm}}{1 \text{ m}}\right) = 34,000 \text{ cm} = 3.4 \times 10^4 \text{ cm}$

(28) 0.00125 g $\qquad$ $(1.25 \text{ mg})\left(\dfrac{1 \text{ g}}{1000 \text{ mg}}\right) = 0.00125 \text{ g}$

(29) 8.9 mg $\qquad$ $(0.0089 \text{ g})\left(\dfrac{1000 \text{ mg}}{1 \text{ g}}\right) = 8.9 \text{ mg}$

(30) 9×10^4 μm $\qquad$ $(9 \text{ cm})\left(\dfrac{1 \text{ m}}{100 \text{ cm}}\right)\left(\dfrac{10^6 \text{ } \mu\text{m}}{\text{m}}\right) = 9 \times 10^4 \text{ } \mu\text{m}$

6. (1) $x = 83$
    (2) $x = 1.4$
    (3) $x = 61$
    (4) $x = 16$

7. (1) We need to change 6 ft 6 in. into inches and then convert to the metric units of centimeters and meters.

$$6 \text{ ft 6 in.} = (6 \text{ ft})\left(\frac{12 \text{ in.}}{\text{ft}}\right) + 6 \text{ in.} = 78 \text{ in.}$$

$$(78 \text{ in.})\left(\frac{2.54 \text{ cm}}{\text{in.}}\right) 2.0 \times 10^2 \text{ cm}$$

$$(2.0 \times 10^2 \text{ cm})\left(\frac{1 \text{ m}}{100 \text{ cm}}\right) = 2.0 \text{ m (2 significant figures)}$$

(2) Since 1 oz. = 29.6 mL, we need to set up the problem to cancel out oz. and give an answer with units of mL.

$$(46.0 \text{ oz.})\left(\frac{29.6 \text{ mL}}{1 \text{ oz.}}\right) = 1.36 \times 10^3 \text{ mL (3 significant figures)}$$

(3) 13¢ per g

We must first convert 1 pound to grams and then divide the cost by the number of grams.

$$(1 \text{ lb})\left(\frac{453.6 \text{ g}}{\text{lb}}\right) = 453.6 \text{ g}$$

59¢/454 g = 13¢ per g (2 significant figures)

8. (1) 0°C   (2) 258K   (3) 238K   (4) 298K
    (5) 75°C   (6) 478K   (7) −244°C
    (8) Low temp °C = (-90.4° − 32)/1.8 = -68.0°C
    High temp °C = (94° − 32)/1.8 = 34°C
    Temp difference = 34°C − (-68.0°C) = 102°C
    (9) °F = (450° x 1.8) + 32 = 842 °C = 840°C

9. (1) 2.7 g/cm³

The mass is given as 2700 g, and we need to determine the volume of a cube 10 cm on a side. Volume is length × width × height. Multiplying 10 cm × 10 cm × 10 cm equals 1000 cm³. Substituting into our equation, we have

$$d = \frac{\text{mass}}{\text{volume}} = \frac{2700 \text{ g}}{1000 \text{ cm}^3} = 2.7 \frac{\text{g}}{\text{cm}^3}$$

The answer says that aluminum metal has a mass of 2.7 g per each cm^3. Therefore, 2 cm^3 would have a mass of 5.4 g, and so forth.

(2) 696 cm^3

We are given the density of the wood and the mass and asked to solve for volume. We write down the equation, then isolate the unknown quantity and take a close look at how the units of a problem can help us in solving equations.

$$d = \frac{\text{mass}}{\text{volume}}$$

Multiply each side by volume.

$$\text{volume} \times d = \frac{\text{mass} \times \cancel{\text{volume}}}{\cancel{\text{volume}}}$$

Now divide each side by density.

$$\frac{\text{volume} \times d}{d} = \frac{\text{mass}}{d}$$

Substitute our values into the modified equation.

$$\text{volume} = 750\,g \times \frac{1\ \text{cm}^3}{1.077\,g}$$

$$= 696\ \text{cm}^3\ \text{(3 significant figures)}$$

Notice that "g" cancels out then leaves our answer with the units for volume, which are the desired units.

10. From the information given, if 1 nm $= 10^{-9}$ m, then 1 m $= 10^9$ nm. Therefore, we can say that

$$1\ \text{m} = 10^9\ \text{nm} = 1,650,763.73\ \text{wavelengths}$$

To find the length of one wavelength we can set up the following two relationships:

$$10^9\ \text{nm} = 1,659,763.73\ \text{wavelengths}$$
$$\frac{10^9\ \text{nm}}{1,650,763.73} = 1\ \text{wavelength (dividing each side by 1,650,763.73)}$$
$$606\ \text{nm} = 1\ \text{wavelength (to three significant figures)}$$

11. (1) 1055 kg: (a) 1055 is the numerical value, kg is the unit (b) 4 significant figures (c) 105 are known, last 5 estimated (d) 1100 kg to 2 significant figures (e) 1.1×10^3 kg

(2) 1,650,763 wavelengths: (a) 1,650,763 is the numerical value, wavelengths are the units (b) 7 significant figures (c) 165076 are known, last 3 is estimated (d) 1,700,000 wavelengths to 2 significant figures (e) 1.7×10^6 wavelengths

(3) 1.077 g/cm^3: (a) 1.077 is the numerical value, g/cm^3 are the units (b) 4 significant figures (c) 1.07 known, last 7 estimated (d) 1.100 g/cm^3 to 2 significant figures (e) 1.1 × 10^0 g/cm^3

(4) 0.00845 L (a) 0.00845 is the numerical value, L is the unit (b) 3 significant figures (c) 84 known, 5 estimated (d) 0.0085 L to 2 significant figures (e) 8.5 × 10^{-3} L

(5) 776,000 g (a) 776,000 is the numerical value, g is unit (b) 3 significant figures (c) 77 known, 6 estimated (d) 780,000 g to two significant figures (e) 7.8 × 10^5 g

CHAPTER THREE

Elements and Compounds

SELECTED CONCEPTS REVISITED

The case of the symbols used is exceedingly important especially if you turn in hand-written work. For example, CA, CA, and ca are unacceptable as a symbol for calcium. The correct symbol is Ca. Using the appropriate sizes and cases (upper versus lower) of letters is crucial.

Many instructors provide you with a periodic table for use during quizzes, tests etc. Although the periodic table shows symbols, most tables do not have the names of the elements. You will most likely have to learn the names of the most common ones. A couple symbols that students seem to get backwards often are P and K. P is phosphorus, K is potassium. The more you use these symbols, the faster you will learn the names.

Being able to identify metals vs. metalloids vs. nonmetals will be invaluable as you learn more about the reactivity of elements and compounds. Remember that the first few chapters of the text form the foundation of your chemistry knowledge.

Elements found in nature are not necessarily found in their elemental form. For example, although gold is often found in a relatively pure state, aluminum is generally found as a part of a compound (or compounds) that must be chemically converted to pure elemental aluminum.

COMMON PITFALLS TO AVOID

It is important that subscripts are written as such and that parenthesis are in the appropriate position relative to the subscripts. $CuCN$ is not the same as $CuCN_2$ which is not the same as $Cu(CN)_2$.

Double-check all chemical formula in both typewritten and handwritten work. When writing or typing Co, did you mean Co for Cobalt, or CO for carbon monoxide? Is PB referring to lead (Pb) or phosphorus and boron? Is MN supposed to be Mn for manganese or MN for a metal-nitrogen compound? Be aware that word-processing programs often automatically change the second letter of a word that has been capitalized back to lower case letters. Spell-check programs do not usually recognize chemical symbols.

SELF-EVALUATION SECTION

1. Fill in the missing word(s) from each sentence.

    The building blocks of all substances are called (1) _____; at least 88 occur naturally on earth.

    (2) _____ are composed of two or more elements chemically combined.

    When an element is subdivided into smaller and smaller pieces, the final indivisible particle that is left is called an (3) _____.

    When a compound, such as sugar, is subdivided into smaller and smaller pieces, the final indivisible particle that retains the identity of sugar, is called a (4) _____.

    Some compounds are formed from electrically charged species called (5) _____.

    Positively charged ions are called (6) _____ and negatively charged ions are called (7) _____.

2. True or False

    (1) Atoms can combine only in simple whole number ratios.

    (2) Ionic compounds can be broken down into molecules.

    (3) There are over 100 known elements.

    (4) A chemical compound retains the identities of its component elements.

3. Fill in the blank spaces.

    All the elements, regardless of origin, can be classified into three subgroups: metals, (1) _____, and (2) _____.

    Metals, except, for mercury, are in the (3) _____ state at room temperature.

    Common characteristics of metals include that they are good conductors of (4) _____ and (5) _____ and usually appear (6) _____.

    Metals are also (7) _____, which means they can be drawn into wires, and are (8) _____, which means they can be flattened into sheets.

    Most metals have (9) _____ melting points and (10) _____ densities.

    Silicon and arsenic are examples of (11) _____.

    Metals tend to react mainly with (12) _____ elements while nonmetals will react with any of the three subgroups.

    There are (13) _____ elements that occur naturally as diatomic molecules and they are (14) _____ (name all). Most of the diatomic elements are gases, but (15) _____ is a liquid and (16) _____ is a solid at room temperature.

4. Write the chemical symbol for the five most abundant elements in the earth's crust, seawater, and atmosphere. Also write the six most abundant elements in the human body. Which elements are common to both lists?

Earth's Chemicals		**Human Body**	
a.	_____	a.	_____
b.	_____	b.	_____
c.	_____	c.	_____
d.	_____	d.	_____
e.	_____	e.	_____
		f.	_____

Elements common to both:

5. Although students are often not asked to memorize the periodic table, knowing the symbols for some of the more common elements will be to the student's advantage. Please write the chemical symbol or name as appropriate for each element below.

Si	_____	Mn	_____
S	_____	Mg	_____
Na	_____	Mercury	_____
N	_____	Gold	_____
Ni	_____	Silver	_____
Phosphorus	_____	Lead	_____
Potassium	_____	Tungsten	_____
C	_____	Tin	_____
Cu	_____	Titanium	_____
Co	_____	B	_____
Calcium	_____	Ba	_____
Cadmium	_____	Br	_____

6. Look at the following list of chemicals. From what you've learned in Chapter 3, try to identify each formula as to whether it represents a compound or not, and if the formula is that of a compound whether the compound is molecular or ionic in nature as described in Chapter 3.

Formula	Compound? Yes or No	If yes then Molecular	or	Ionic
(1) Br_2	_____		_____	_____
(2) CO_2	_____		_____	_____
(3) S_8	_____		_____	_____
(4) NH_3	_____		_____	_____
(5) H_2O	_____		_____	_____
(6) CCl_4	_____		_____	_____
(7) Pb	_____		_____	_____
(8) KCl	_____		_____	_____
(9) Ar	_____		_____	_____
(10) NaI	_____		_____	_____

7. How many atoms of nitrogen are contained in each of the following formulas?

(1) $Na_3Co(NO_2)_6$ _____

(2) $NH_4S_2O_8$ _____

(3) $NH_4C_2H_3O_2$ _____

(4) NH_4CNS _____

(5) $Na_2Fe(CN)_5NO \cdot 2\ H_2O$ _____

(6) $K_3Fe(CN)_6$ _____

(7) $Fe_2(SO_4)_3(NH_4)_2SO_4 \cdot 24\ H_2O$ _____

RECAP SECTION

Chapter 3 has given you an introduction to the vocabulary of the chemist. We have classified matter into substances and mixtures, then distinguished between elements and compounds. The new terms of atom, molecule, cation, and anion were also added to our vocabulary. We have learned to identify the physical states of matter, to distinguish among metals, nonmetals, and metalloids, and to understand the meaning of a chemical formula. With this new understanding of the language of chemistry you are ready to study the properties of matter.

ANSWERS TO QUESTIONS AND SOLUTIONS TO PROBLEMS

1. (1) elements   (2) Compounds   (3) atom
    (4) molecule   (5) ions   (6) cations
    (7) anions

2. (1) True   (2) False   (3) True   (4) False

3. (1), (2) metalloids, nonmetals   (3) solid
    (4), (5) heat, electricity   (6) lustrous or shiny
    (7) ductile   (8) malleable   (9) very high
    (10) high   (11) metalloids   (12) nonmetal
    (13) seven   (14) O_2, N_2, F_2, Cl_2, O_2, Br_2, I_2
    (15) bromine   (16) iodine

4. Earth: O, Si, Al, Fe, Ca
    Body: O, C, H, N, Ca, P
    Oxygen and calcium are common to both.

5. Si - silicon   Mn - manganese
    S - sulfur   Mg - magnesium
    Na - sodium   Mercury - Hg
    N - nitrogen   Gold - Au
    Ni - nickel   Silver - Ag
    Phosphorus - P   Lead - Pb
    Potassium - K   Tungsten - W
    C - carbon   Tin - Sn
    Cu - copper   Titanium - Ti
    Co - cobalt   B - boron
    Calcium - Ca   Ba - barium
    Cadmium - Cd   Br - bromine

6. (1) No   (2) Yes, molecular   (3) No   (4) Yes, molecular   (5) Yes, molecular   (6) Yes, molecular   (7) No
    (8) Yes, ionic   (9) No   (10) Yes, molecular

7. (1) 6   (2) 1   (3) 1   (4) 2   (5) 6   (6) 6   (7) 2

CHAPTER FOUR

Properties of Matter

SELECTED CONCEPTS REVISITED

A chemical change results in a new substance. Often you will see physical changes accompanying a chemical change because the new substance will likely have different physical characteristics than the original compound. If only a physical change takes place, no new substance is formed.

A few reminders about chemical equations (they describe chemical reactions)…

reactants	$\rightarrow$	products	
(starting substances	$\rightarrow$	new substance(s) formed)	
(total mass of reactants	=	total mass of products)	*Law of Conservation of Mass*

Potential energy is the energy which matter has as a result of chemical bonds or because of its position in relationship to another place; kinetic energy is strictly the energy of motion.

When we talk about the amount of heat energy involved in a chemical process, we use the SI unit of measurement called the *joule*. We define the joule by an experimental procedure – a common practice in science. If we measure out 1 gram of water and place it at 14.5° Celsius, then 4.184 joules of heat energy from a match, gas burner, or other source will raise the temperature of the 1 gram of water 1° Celsius to 15.5° Celsius. In words, the number of joules gained or lost is equal to the mass of water times the temperature change times a constant. The constant is the *specific heat* for the material in question. The higher the specific heat, the more heat is needed to raise the temperature by one degree. Therefore, a substance with higher specific heat will need to absorb more energy to raise its temperature than a substance with a lower specific heat. For water, the specific heat is 4.184 J/g°C.

Remembering that the number of joules (abbreviated J) gained or lost during heat energy changes is equal to the mass of the substance times the specific heat of the substance times the temperature difference, we can write the following equation.

$$\text{joules} = (\text{grams of substance})(\text{specific heat of substance})(\Delta\ t)$$

The symbol $\Delta\ t$ means "temperature difference". The next to the last term in the equation is the specific heat for the substance in question. For water, the specific heat is 4.184 J/g°C. An equation with units attached would be

$$\text{joules} = \text{g} \times 4.184\ \text{J/g°C} \times \text{°C}$$

You have learned that heat is not the same as temperature. They are however inextricably linked. Heat flows between two bodies until the temperatures are equal. Therefore bodies at higher temperatures lose heat until they reach the same temperature as a body at a cooler temperature. This is not to say that a body at high temperature necessarily has more heat energy than a body at a cooler temperature since the amount of heat energy within the body is related to its mass.

COMMON PITFALLS TO AVOID

Do not forget to be consistent with your units! One of the most common mistakes students make is to forget to check that their units are consistent. You cannot add a number in Joules to a number in kilojoules without first converting one of the numbers to the same unit as the other.

It is best not to start to solve a problem without first reading through the entire problem. Here are some hints to help you solve problems. Remember when you approach problem solving you should always read the entire problem through once to get a quick feel for the topic and what you are being asked to solve. You should then reread the problem picking out the important information and organizing it in a way you can easily understand. For example, if a problem involves numbers such as "the mass of an object occupying 12 ml is 30.5 g", write the variable symbol followed by the given number for easy reference (e.g. m = 30.5 g, V = 12 ml). Try to determine what you are being asked for and where you can get the information you need, that is, is it given to you in the question, is it something you are supposed to have memorized, is it found on the periodic chart, etc?

Heat $\neq$ temperature; heat content depends on the mass of the object. Temperature is independent of mass.

SELF-EVALUATION SECTION

1. Statements about the properties and changes of various substances are given below. Place the appropriate letter of identification in each space provided.

    a. physical property   b. chemical property
    c. physical change   d. chemical change

Description	**Identification**
(1) Potassium is solid at room temperature.	_____
(2) Potassium is among the most reactive elements in nature.	_____
(3) The melting point of potassium is 63.6 °C.	_____
(4) Potassium reacts vigorously with water.	_____
(5) Potassium added to water produces hydrogen gas plus a water solution of potassium hydroxide.	_____
(6) At 760 °C, potassium begins to boil and to change from a liquid to a gas.	_____
(7) At high temperatures, potassium reacts with hydrogen gas to form a metallic hydride.	_____

Description	Identification

(8) Alkali metals, such as potassium, form salts
with many other elements. _____

(9) Potassium metal is shiny and fairly soft. _____

(10) Potassium is useful in the manufacture of fertilizer
compounds. _____

2. Determine whether the following statements or blanks relate to *chemical, light, heat, mechanical,* or *electrical* energy.

Radiant energy __(1)__ from the sun is used in the (1) _____
process of photosynthesis.
Photosynthesis involves a conversion of radiant
energy into the __(2)__ energy of sugars, starch,
and other compounds. (2) _____
Organisms utilize the __(3)__ energy of food-
stuffs for movement and activity (3) _____
and, in the case of warm-blooded animals, for
maintaining body temperature __(4)__ . (4) _____
When wood is burned, the two forms of energy (5) _____
that are released are __(5)__ and __(6)__ . (6) _____
The chemical energy of a car battery is converted
into __(7)__ energy, which turns the (7) _____
starter motor.
Fuels such as gasoline contain __(8)__ energy (8) _____
which, through the process of combustion, is
converted to the __(9)__ energy of moving pistons, (9) _____
valves, gears, and wheels.
Hydroelectric dams use the energy of falling
water to turn turbines that convert __(10)__ (10) _____
energy into __(11)__ energy. (11) _____

3. The specific heats of aluminum, iron and copper are 0.900, 0.473, and 0.385 J/g°C respectively. Three identical lawn statues were taken out of a room at 25°C and placed together in the sun for a half hour. Which of the statues would have the highest temperature? Explain briefly.

4.   (1) How many kilojoules of heat energy would be required to heat a beaker containing 250 mL (250 g) of water from room temperature (25°C) to 95°C?

Do your calculations here.

(2) How many grams of water can be heated from 5 °C to 52 °C with 1.0×10^2 joules of heat energy?

Do your calculations here.

(3) Let's work problems (1) and (2) again but this time we will use the calorie unit of heat measurement. Recall that the specific heat of water in calorie units is 1 cal/g°C or in words "one calorie of heat energy will raise one gram of water one degree Celsius". Therefore, how many kilocalories of heat energy are required to heat 250 g of water from 25 °C to 95 °C?

Do your calculations here.

(4) In a similar manner, how many grams of water can be heated from 5 °C to 52 °C with 1.0×10^2 cal of heat energy?

Do your calculations here.

Challenge Problems

5. Under the right conditions water can be made from hydrogen gas and oxygen gas.

$$2\ H_2 + O_2 \rightarrow 2\ H_2O$$

   (1) What is (are) the reactant(s)? What is (are) the product(s)?

   (2) If at the start of the reaction, 20.0 g each of H_2 and O_2 are used, and 17.5 g of H_2 is left when the reaction is complete, what is the percent composition of H_2 and O_2 in H_2O?

   Do your calculations here.

   (3) Which law did you apply to help solve part (2)?

6. A 50 g piece of copper at 212 °F is dropped into 100 g of ethyl alcohol at 72 °F. The specific heat of copper is $0.0921\ \dfrac{cal}{g°C}$, that of ethyl alcohol is $0.511\ \dfrac{cal}{g°C}$. Calculate the final temperature of the mixture.

   Do your calculations here.

7. Three substances A, B, and C are undergoing investigation in the laboratory. The specific heat capacity of A is known to be 0.517 J/g°C.
   (1) 10.0 g each of A and B are heated under identical conditions. After 5 minutes of heating, A is 3 °C warmer than B. Which substance has a higher specific heat, A or B? Briefly explain.
   (2) The specific heat of C is found to be nearly twice that of A. Will 5.0 g of C or 5.0 g of A lose more energy when each sample drops 8 °C? Briefly explain.

RECAP SECTION

In Chapter 4 we examined the physical and chemical properties of matter. Physical properties can be determined without altering the substance, while chemical properties involve the interaction of substances resulting in different material. Whether matter undergoes a physical change or a chemical change, energy is conserved. Energy can take many forms, including the heat and light most commonly seen during chemical changes. The material in these first chapters has given you the foundation to begin to explore the atom and its internal structure.

ANSWERS TO QUESTIONS AND SOLUTIONS TO PROBLEMS

1. (1) a   (2) b   (3) a   (4) b
    (5) d   (6) c   (7) d   (8) b
    (9) a   (10) b

2. (1) light   (2) chemical   (3) mechanical   (4) heat
    (5) heat   (6) light   (7) electrical   (8) chemical
    (9) mechanical   (10) mechanical   (11) electrical

3. The copper statue.
    Assuming all three absorbed the same amount of heat, the biggest change in statue temperature should come from the statue with the lowest specific heat.

4. (1) 73 kJ. We know the mass of water (250 g) and can determine the temperature difference by subtraction. The unknown quantity is the number of kilojoules of heat energy required. Specific heat has units involving joules so the heat energy is first found in joules and then divided by 1000 to give the answer in kilojoules (and two significant figures).

$$
\begin{aligned}
\text{joules} &= m \text{ x specific heat x } \Delta T \\
&= (250 \text{ g})(4.184 \text{ J/g°C})(95\text{-}25\text{°C}) \\
&= 73220 \text{ J} \\
\text{kJ} &= 73 \text{ kJ}
\end{aligned}
$$

   (2) 5.1 g of water (2 significant figures)
        We are asked to determine the mass of water which 1.0×10^2 joules of heat energy can raise from 5 °C to 52 °C. Solving our equation for grams we have

$$
\begin{aligned}
\text{joules} &= m \times \text{specific heat} \times \Delta t \\
m &= \frac{\text{joules}}{(\text{specific heat})(\Delta t)} \\
&= \frac{1.0 \times 10^2 \text{ joules}}{(4.184 \text{ J/g°C})(52 \text{ °C} - 5\text{°C})} \\
&= \frac{1000 \text{ joules}}{(4.184 \text{ J/g°C})47\text{°C}} \\
&= 5.1 \text{ g of water}
\end{aligned}
$$

(3) 18 kcal to 2 significant figures

We know the mass of water (250 g) and know the temperature difference by subtraction. The unknown quantity again is the number of calories of heat energy required. Make the necessary substitutions in the equation.

$$cal = (250 \text{ g})(1 \text{ cal/g°C})(95 \text{ °C} - 25 \text{ °C})$$
$$= (250 \text{ g})(1 \text{ cal/g°C})(70\text{°C})$$
$$= 17{,}500 \text{ cal}$$
$$kcal = 18 \text{ kcal (2 significant figures)}$$

(4) 21 g of water

Again, what mass of water can 1.0×10^2 cal raise from 5 °C to 52 °C?

Solving our equation for grams we will have

$$cal = m \times \text{specific heat} \times \Delta t$$

$$m = \frac{cal}{(\text{specific heat})(\Delta t)}$$

$$= \frac{1.0 \times 10^2 \text{ cal}}{(1 \text{ cal/g°C})(52 \text{ °C} - 5 \text{ °C})}$$

$$= \frac{1.0 \times 10^2 \text{ cal}}{(1 \text{ cal/g°C})(47 \text{ °C})}$$

$$= 21 \text{ g of water}$$

5. (1) reactants are H_2 and O_2; product is H_2O

(2) 11.1% H_2 and 88.9% O_2

From the data given, the mass of water formed is the amount of O_2 present plus the amount of H_2 reacted.

$$20.0 \text{ g } O_2 + 2.5 \text{ g } H_2 = 22.5 \text{ g } H_2O$$

$$\% \, H_2 = \left(\frac{\text{mass } H_2}{\text{mass } H_2O}\right)100 = \left(\frac{2.5 \text{ g}}{22.5 \text{ g}}\right)100 = 11.1\%$$

$$\% \, O_2 = \left(\frac{\text{mass } O_2}{\text{mass } H_2O}\right)100 = \left(\frac{20.0 \text{ g}}{22.5 \text{ g}}\right)100 = 88.9\%$$

(3) Law of Conservation of Mass

6. In examining the problem we should notice first that the temperatures are given in °F while specific heats contain the temperature unit of °C. In general, as we work with chemical problems we will find that conversions are necessary when we see temperatures given as °F. Therefore, the copper metal's temperature is

$$°C = \frac{(°F - 32)}{1.8} = \frac{(212 - 32)}{1.8} = 100 \text{ °C}$$

The temperature of the ethyl alcohol is

$$°C = \frac{(°F - 32)}{1.8} = \frac{(72 - 32)}{1.8} = 22 \text{ °C}$$

Next we can say that the copper metal piece has been heated up, dropped into a cooler material, the alcohol, and caused the final mixture to increase in temperature. The calories of heat gained by the copper as it was heated are lost to the alcohol so that the final temperature of the copper and alcohol are equal.

Another way to state this is that the heat lost by the copper is equal to the heat gained by the alcohol.

Mathematically, the equality would be:

$$\text{heat lost}_{Cu} = \text{heat gained}_{alcohol}$$

$$\text{heat lost}_{Cu} = m_{Cu}c_{Cu}\,\Delta\,T_{Cu} = (50\text{ g})\left(0.092\,\frac{cal}{g°C}\right)(100\text{ °C} - T_f)$$

$$\text{Heat gained}_{alcohol} = m_{alcohol}c_{alcohol}\,\Delta\,T_{alcohol} = (100\text{ g})\left(0.511\,\frac{cal}{g°C}\right)(T_f - 22\text{ °C})$$

(c_{Cu} and $c_{alcohol}$ represent the specific heats of copper and alcohol respectively.)
Notice that T_f is second in $\Delta\,T_{Cu}$ and first in $\Delta\,T_{alcohol}$. This is a result of the Cu being cooled (to a lower temperature) while the alcohol is being warmed (to a higher temperature).

Since heat lost$_{Cu}$ = heat gained$_{alcohol}$

$$(50\text{ g})\left(0.0921\,\frac{cal}{g°C}\right)(100\text{ °C} - T_f) = (100\text{ g})\left(0.511\,\frac{cal}{g°C}\right)(T_f - 22\text{ °C})$$

Upon expansion,

$$460.5\text{ cal} - \left(4.605\,\frac{cal}{°C}\right)T_f = \left(51.1\,\frac{cal}{°C}\right)T_f - 1124.2\text{ cal}$$

$$1584.7\text{ cal} = \left(55.705\,\frac{cal}{°C}\right)T_f$$

$$28\text{ °C} = T_f$$

7.   (1) B has a higher specific heat than A. Substance B required more energy to raise its temperature by a degree so for the same amount of energy input, its overall temperature rise was smaller than that of substance A.
     (2) Substance C will lose more energy than substance A. If C has a higher specific heat it means more energy is involved in changing the temperature of a sample by one degree. Therefore for similar temperature drops, C must have released nearly twice the amount of energy than A.

CHAPTER FIVE

Early Atomic Theory and Structure

SELECTED CONCEPTS REVISITED

The Law of Definite Composition states that a compound has two or more elements in a definite proportion by mass. The Law of Multiple Proportions states that atoms of two or more elements may combine in different ratios to produce more than one compound.

Two important features of Dalton's theory were later proved wrong by the existence of isotopes and by the discovery of subatomic particles. Atoms are organized as a small nucleus containing the protons and the neutrons. A cloud of electrons surrounds the positively charged nucleus. Electrons are the only subatomic particles involved in most reactions with the exception of nuclear reactions (you will cover nuclear reactions in a later chapter). When electrons are lost or gained from an atom, the atom becomes charged and is referred to as an ion (a charged species). Because the protons and the neutrons have much greater mass than the electrons, we generally do not distinguish between the mass of an ion versus the mass of the corresponding atom.

The atomic mass unit (amu) is introduced in this chapter. The weight of a single atom is very small so scientists developed a new unit (amu) to avoid having to use numbers on the order of 10^{-24} on a regular basis. The numbers for the masses on the periodic chart when followed by the unit amu refer to the weight of single atoms. You will see in a later chapter that when grams are used with the periodic chart numbers, many more than single atoms are being counted/weighed. Amu's are not weighed in the lab.

A law describes behavior; a theory explains behavior.

COMMON PITFALLS TO AVOID

Try not to write charges unnecessarily. It is important to know when it is appropriate to write the charge on an ion. When an ion is written by itself as opposed to as part of a compound, the charge must be shown. The charge is written as a superscript to the right of the symbol. Na is not the same as Na^+. When the ion is part of a compound the charges are generally not written, for example, we usually write NaCl not Na^+Cl^- (you may see exceptions to this used when we want to emphasize the charges).

If the charge on an ion is greater than 1, write the number before the sign, e.g., Mg^{2+} not Mg^{+2}. You will later learn that the number after the sign can mean something else (oxidation number).

Isotopes must have the same atomic number. $^{14}_{6}C$ and $^{14}_{7}N$ are not isotopes.

1.   Match the scientist to the proposal, discovery, or theory he made.

   Chadwick   ___ (1) discovered radioactivity

   Becquerel   ___ (2) alpha particle experiment results in proposal of a heavy, dense atomic nucleus

   Thomson   ___ (3) discovered the neutron, a particle with mass but no charge

   Rutherford   ___ (4) experimentally showed the existence of the atom; also discovered proton are particles

   As a result of these and other experiments and discoveries, the three sub-atomic particles of interest are (5) _____, (6) _____, and (7) _____.

   This also led to two subsequent proposals for a model of an atom as illustrated below.
   (8) Which model was proposed first and by whom?

2.   Atoms of a particular element which have the same atomic (1) _____ but differ in atomic mass are called (2) _____ of that element. Some radioactive isotopes are extremely important today. Some such as $^{131}_{53}I$, $^{60}_{27}Cu$, or $^{32}_{15}P$ are used in medical applications. Others, such as $^{90}_{38}Sr$, are environmental contaminants and may pose a long-term problem for various organisms. The simplest element, hydrogen, has three isotopes. They are named (3) _____, (4) _____, and (5) _____ and are represented as (6) _____, (7) _____, and (8) _____.

3.   Indicate by a yes or no response which of the following statements are generally accepted today.
   (1)   Atoms of the same element may vary in mass but are the same size. _____
   (2)   Compounds are formed by the union or joining of two or more atoms of different element. _____
   (3)   Atoms of different elements have different masses and sizes. _____
   (4)   Atoms of two elements may combine in different ratios to form more than one compound. _____
   (5)   Elements are composed of small, indivisible units called electrons. _____
   (6)   In forming compounds, atoms can combine in fractional units. _____
   (7)   The fundamental building blocks of elements are small particles called atoms. _____
   (8)   Atoms of the same element are alike in mass and size. _____
   (9)   In forming compounds, atoms combine in small whole-number ratios such as one to one, one to two, and so forth. _____
   (10) Atoms of two elements can combine in only one fixed ratio to form only one compound. _____

4. Fill in the blank spaces.

(1) _____ number ⟶ ^3_1H
(2) _____ number ⟶

(1) _____ number = number of (3)_____ + number of (4)_____

(2) _____ number = number of (5)_____

In a neutral atom, the number of (6)_____ = (7)_____

5. Description of properties of subatomic particles. Fill in the spaces of the table with the correct response from the list of possible responses below.

Particle	Symbol	Relative Mass (amu)	Charge	Location in Atom
Electron				
Proton				
Neutron				

Possible responses

Symbol	Relative Mass (amu)	Charge	Location in Atom
	$+2$	-1	nucleus
p	$+1$	-2	outside nucleus
	-1	$+1$	
e	$-\frac{1}{2}$	-2	
n	0	0	
	$\frac{1}{1837}$	$-\frac{1}{2}$	

6. Atomic masses, atomic numbers, and number of neutrons.
   (1) Determine the number of neutrons in an atom of each of the following elements.

   $^{23}_{11}\text{Na}$ _____ $^{65}_{30}\text{Zn}$ _____

   $^{115}_{48}\text{Cd}$ _____ $^{40}_{18}\text{Ar}$ _____

   $^{37}_{17}\text{Cl}$ _____ $^{24}_{12}\text{Mg}$ _____

(2)   Determine the approximate *atomic mass* for an atom of each of the following isotopes.

	Atomic Number	Number of Neutrons	Atomic Mass
Li	3	4	_____
Al	13	14	_____
Bv	35	46	_____
Fe	26	30	_____
H	1	0	_____
O	8	10	_____

7.   Which of the following statements are true about the neutral atoms $^{12}_{6}C$ and $^{14}_{6}C$?
    (1) $^{12}_{6}C$ and $^{14}_{6}C$ are isotopes.
    (2) $^{12}_{6}C$ has two less electrons than $^{14}_{6}C$.
    (3) $^{14}_{6}C$ has two more protons than $^{12}_{6}C$.
    (4) $^{12}_{6}C$ has two less neutrons than $^{14}_{6}C$.

8.   a)   The actual mass of a Carbon-12 atom is 1.9927×10^{-23} g. What is the relative atomic mass (i.e., the mass that would appear under the symbol of the element on the periodic table) of an element Jamcium whose mass is 4.0986×10^{-21}?
      b)   If there are only two isotopes of this element and they differ by only one neutron, which is more naturally abundant, the lighter or the heavier isotope? How do you know?

RECAP SECTION

Chapter 5 provides our first look inside the atom. Early models of the atom were modified to incorporate the concept of electric charge. The atom contains three types of subatomic particles, the proton, the neutron, and the electron. The arrangement of these particles within the atom was proposed by Ernest Rutherford. In his model of the atom the neutrons and the protons are located within the nucleus, while the electrons occupy the remainder of the atom. This brief glimpse into atomic structure allows us to relate the atomic number to the number of protons in the nucleus (they are equal). These early models of the atom also led to the discovery that isotopes are atoms of the same element with differing numbers of neutrons. From here we can begin to look at the composition of compounds.

ANSWERS TO QUESTIONS AND SOLUTIONS TO PROBLEMS

1.   (1) Becquerel   (2) Rutherford   (3) Chadwick   (4) Thomson
      (5), (6), (7) proton, neutron, electron (in any order)
      (8) Thomson proposed the model on the right first. After his gold foil experiment, Rutherford proposed the heavy, dense nucleus model of the atom.

2.   (1) number   (2) isotopes
      (3), (4), (5) protium, deuterium, tritium
      (6), (7), (8) $^{1}_{1}H$, $^{2}_{1}H$, $^{3}_{1}H$

3. (1) yes   (2) yes   (3) yes   (4) yes   (5) no
    (6) no   (7) yes   (8) no   (9) yes   (10) no

4. (1) mass   (2) atomic   (3), (4) protons, neutrons
    (5) protons   (6), (7) protons, electrons

5.

Particle	Symbol	Relative Mass (amu)	Charge	Location in Atom
Electron	e	1/1837	-1	outside nucleus
Proton	p	1	$+1$	nucleus
Neutron	n	1	0	nucleus

6. (1)   Na = 12   Zn = 35
           Cd = 67   Ar = 22
           Cl = 20   Mg = 12
    (2)   Li = 7   Fe = 56
           Al = 27   H = 1
           Br = 81   O = 18

7. (1) T   (2) F   (3) F   (4) T

8. a) 205.68
    b) The heavier isotope. Since the isotopes differ by only one mass number, the nuclides must weigh 205 and 206. If there were equal amounts of each isotope, the average mass would be 205.5. Since the average mass is greater than 205.5, the heavier isotope must be more abundant.

CHAPTER SIX

Nomenclature of Inorganic Compounds

SELECTED CONCEPTS REVISITED

Naming compounds is an essential aspect of chemistry. Some names and their corresponding chemical formula you will simply have to memorize, such as the common names for certain compounds. Ask your instructor for the specific common names that will be required to know. Some of the more popular ones are laughing gas, N_2O, baking soda, $NaHCO_3$, and table salt. You will also need to become familiar with many polyatomic ions. The oxygen-containing polyatomic ions often involve the root name of the non-oxygen component, for example, phosphate would contain P and O (PO_4^{3-}), and nitrite contains N and O (NO_2^-). As a helpful hint, the more you use the names and formula, the faster you will learn them. Get into the habit of trying to name all simple compounds you meet in the text or in class.

For ionic compounds, the name consists simply of the name of the cation (using a Roman numeral if necessary) followed by the name of the anion.
For covalent compounds it is important to remember to use prefixes to indicate the formula of the compound.
Compounds that are acids (the formula begins with H) can be named either as ionic compounds or specifically as acids.

The subscript in a formula indicates the number of atoms chemically bonded together to form that molecule or unit of a compound. Roman numerals are used only in the names of transition metal compounds and not in their formula.

Although most transition metals have more than one possible ion, it is important to understand in a particular compound only one ion exists at a time. That is, if you broke down the compound Mn_2O_5 both manganese ions are Mn^{5+} (*not* one Mn^{7+} and one Mn^{3+}).

Parentheses are necessary to show more than one polyatomic ion in a formula.

COMMON PITFALLS TO AVOID

The subscript on a transition metal is not the same as the Roman numeral after the metal's name in the compound name. The subscript indicates the number of ions involved in a single unit of the compound while the Roman numeral indicates the *charge* on the metal ion. Roman numerals are only necessary when the transition metal has more than one possible ion.

Roman numerals are <u>not</u> written in the chemical formula of a compound.

Be careful when referring to ions, for example, do *not* refer to K^+ as a potassium *atom*. While this may seem to be nit picking, the behavior of a potassium atom is actually very different from the behavior of a potassium ion so it is very important that you refer to the different species correctly.

Trying to name compounds by simply naming every element is a common mistake made in early chemistry classes. Most compounds you meet at this level that are composed of more than two elements will most likely be ionic compounds and involve a polyatomic ion. You will not be able to name ionic compounds if you cannot recognize polyatomic ions. For example, K_2CO_3 is *not* potassium carbon oxide but is potassium carbonate.

Do not try to memorize the formula for ionic compounds. There are too many variations. You need to know the polyatomic ions and you should be able to look at the periodic table to determine the charge of the ions formed from certain elements. Once you know these you should be able to figure out the formula for the different ionic compounds. If you understand this it may make it easier for you to remember that you do not need prefixes (di, tri etc) in the names of ionic compounds to describe the subscripts because we can always determine the subscripts from the charges on the ions. In covalent compounds, if we are not told the subscripts in the name, we have no way of knowing the subscripts in the formula because the formula for the compound is not based on balancing charges.

Try not to memorize all the charges on monatomic ions. When you get to electronic structure of atoms you will be able to understand why certain atoms form specific ions but in the meantime use the periodic chart to help you. The text explains the trends. Remember you generally have access to a periodic chart.

SELF-EVALUATION SECTION

1.   Write formula for the following elements and ions. Recall that the charges must be written for the ions.

   (1) sulfur _____   (2) sulfide _____

   (3) lithium _____   (4) lithium ion _____

   (5) phosphorus _____   (6) phosphide _____

2.   Write the chemical equation showing the loss or gain of electrons for the elements in number 1 above.

3.   Write the correct formula for the ions first and then write the formula for the compound. (These are the steps you should always take when given the name of an ionic compound and asked to write the formula.)

   (1) silver ion   _____   chloride   _____   silver chloride   _____

   (2) aluminum ion   _____   oxide   _____   aluminum oxide   _____

   (3) iron (III) ion   _____   bromide   _____   iron (III) bromide   _____

   (4) ammonium ion   _____   sulfate   _____   ammonium sulfate   _____

   (5) magnesium ion   _____   phosphate   _____   magnesium phosphate   _____

   (6) copper (II) ion   _____   cyanide   _____   copper (II) cyanide   _____

   (7) potassium ion   _____   carbonate   _____   potassium carbonate   _____

   (8) lead (II) ion   _____   nitrate   _____   lead (II) nitrate   _____

4. On the following list of formula,

(1) Circle the acids   (2) Underline the non-metallic binary compounds

HCl   CO SnO_2 Al_2O_3 PBr_3 $PbCl_2$

P_2O_5 H_2CO_3 NaCl $MgCl_2$ H_2SO_4 N_2O_3

5. Write the names for the formulas and the formulas for the written names listed below. Refer to the tables in Chapter 6.

(1)   AgCl   _____

(2) KNO_3 _____

(3) $SnCl_4$ _____

(4) Al_2O_3 _____

(5) SO_3 _____

(6) NH_4NO_3 _____

(7) $AlPO_4$ _____

(8) Na_2SO_4 _____

(9) $(NH_4)_2CO_3$ _____

(10)   NaI   _____

(11) $FeCl_2$ _____

(12)   BaO   _____

(13) $K_2Cr_2O_7$ _____

(14) $NaHCO_3$ _____

(15)   CO   _____

(16)   Hydrochloric acid   _____

(17)   Lead (II) oxide   _____

(18)   Potassium permanganate   _____

(19)   Magnesium chloride   _____

(20) Carbon dioxide _____

(21) Barium sulfate _____

(22) Iron (III) Chloride _____

(23) Silicon dioxide _____

(24) Sulfuric acid _____

(25) Dinitrogen tetroxide _____

(26) Nitric acid _____

(27) Cobalt (II) chloride _____

(28) Sodium sulfite _____

(29) Hydrosulfuric acid _____

(30) Carbon tetrachloride _____

6.   Give the formulas for the following salts, which contain more than one positive ion.

(1)   Sodium hydrogen carbonate

(2)   Magnesium ammonium phosphate

(3)   Sodium potassium sulfate

(4)   Ammonium hydrogen sulfide

(5)   Potassium hydrogen sulfate

(6)   Potassium aluminum sulfate

7.   With which of the following ions can the magnesium ion combine to form a compound of formula MgZ?

(1) sulfide ion   (2) cyanide ion
(3) bromide ion   (4) carbonate ion
(5) nitrate ion   (6) iron (II) ion

8. Write the correct formulas for compounds formed by matching each cation with all anions.

	Cl^-	O^{2-}	NO_3^-	PO_4^{3-}
K^+				
Mg^{2+}				
Fe^{3+}				
H^+				
Cu^+				
Sn^{4+}				
Ca^{2+}				
Al^{3+}				

RECAP SECTION

Chapter 6 is one of the self-contained study units in the text for you to refer to as needed. After completing the material in the text, you will be familiar with naming binary compounds composed of a metal and a nonmetal and those composed of two nonmetals. It is common to find oxygen-containing radicals in polyatomic compounds. Rules are given for naming these compounds containing polyatomic ions. Only through practice and experience in a laboratory will you become adept at using chemical names. Try to use names whenever possible. Ask your instructor the name of any chemical you are not sure about. It is a good practice to have the original chemical bottles present during a lab period even though your instructor may provide the chemicals in solution. Ask to see the reagent bottles and read labels carefully.

ANSWERS TO QUESTIONS

1. (1) S (2) S^{2-} (3) Li (4) Li^+ (5) P (6) P^{3-}

2. Recall that electrons that are lost are written as products (chemical equations do not "minus" substances).
    $$S + 2e^- \rightarrow S^{2-}; \ Li \rightarrow Li^+ + e^-; \ P + 3e^- \rightarrow P^{3-}$$

3. (1) Ag^+, Cl^-, AgCl   (2) Al^{3+}, O^{2-}, Al_2O_3
    (3) Fe^{3+}, Br^-, $FeBr_3$ (4) NH_4^+, SO_4^{2-}, $(NH_4)_2SO_4$
    (5) Mg^{2+}, PO_4^{3-}, $Mg_3(PO_4)_2$ (6) Cu^{2+}, CN^-, $Cu(CN)_2$
    (7) K^+, CO_3^{2-}, K_2CO_3 (8) Pb^{2+}, NO_3^-, $Pb(NO_3)_2$

4. (HCl)　　　　CO　　　　SnO_2　　　　Al_2O_3　　　　PBr₃　　　　$PbCl_2$

　　$MgCl_2$　　(H₂CO₃)　　NaCl　　　　SO₃　　　　(H₂SO₄)　　N_2O_3

5. (1) Silver chloride   (16) HCl
    (2) Potassium nitrate   (17) PbO
    (3) Tin (IV) chloride   (18) $KMnO_4$
    (4) Aluminum oxide   (19) $MgCl_2$
    (5) Sulfur trioxide   (20) CO_2
    (6) Ammonium nitrate   (21) $BaSO_4$
    (7) Aluminum phosphate   (22) $FeCl_3$
    (8) Sodium sulfate   (23) SiO_2
    (9) Ammonium carbonate   (24) H_2SO_4
    (10) Sodium iodide   (25) N_2O_4
    (11) Iron (III) chloride   (26) HNO_3
    (12) Barium oxide   (27) $CoCl_2$
    (13) Potassium dichromate   (28) Na_2SO_3
    (14) Sodium hydrogen carbonate (29) H_2S
    (15) Carbon monoxide   (30) CCl_4

6. (1) $NaHCO_3$ (2) $MgNH_4PO_4$ (3) $NaKSO_4$
    (4) NH_4HS (5) $KHSO_4$ (6) $KAl(SO_4)_2$

7. (1) sulfide ion   (4) carbonate ion

8. KCl K_2O KNO_3 K_3PO_4
    $MgCl_2$ MgO $Mg(NO_3)_2$ $Mg_3(PO_4)_2$
    $FeCl_3$ Fe_2O_3 $Fe(NO_3)_3$ $FePO_4$
    HCl H_2O HNO_3 H_3PO_4
    CuCl Cu_2O $CuNO_3$ Cu_3PO_4
    $SnCl_4$ SnO_2 $Sn(NO_3)_4$ $Sn_3(PO_4)_4$
    $CaCl_2$ CaO $Ca(NO_3)_2$ $Ca_3(PO_4)_2$
    $AlCl_3$ Al_2O_3 $Al(NO_3)_3$ $AlPO_4$

CHAPTER SEVEN

Quantitative Composition of Compounds

SELECTED CONCEPTS REVISITED

The mole has become the center of your universe. Not quite, but close. The mole is simply a name given to a very large number, 6.02×10^{23} (also known as Avogadro's number). A dozen is 12, a gross is 144, a mole is 6.02×10^{23}.

The molar mass of a substance is the mass in grams of 1 mole of particles of that substance. That is, if you took 6.02×10^{23} atoms of gold, put it on a scale and weighed it in grams, that would be the molar mass of gold. Numerically that would be equivalent to the atomic mass of gold. So the atomic mass of gold is 197.0, therefore, 1 mole of gold weighs 197.0 g. That is, the molar mass of gold is 197.0 g/mole.

The empirical formula for a compound is the minimum whole number ratio of each element in a single unit or molecule of that compound. If the relationship between the elements is given as a mass ratio, you need to take into account that atoms of each element have different masses. Therefore the masses should first be converted to moles and since moles represent a numerical quantity, the number of atoms of each element can then be compared.

COMMON PITFALLS TO AVOID

Remember that percent composition refers to the percent by mass and do not forget to take into account the numbers of each type of atom present when calculating percent composition. Thus the percent composition of C in C_2H_6 is

$$\% \text{ comp} = (100)(\text{mass of C in } C_2H_6)/\text{mass of } C_2H_6 = \frac{(2)(12)(100)}{(2)(12) + (6)(1)} = 80\%$$

Do not forget to double-check your answer to see if it makes sense. If you have difficulty realizing whether you should multiply or divide by the molar mass to convert from grams to moles, do a quick qualitative check. For example, you want to convert 135 g of unknown A to moles of A and the molar mass of A is 154 g. Without even doing the actual calculation, you can see that one mole weighs 154 g so if you have less than 154 g you should have less than one mole.

When taking masses from the periodic chart, be consistent about the number of significant digits used. Rarely should the mass from the periodic chart limit the significant figures in the calculation. Also, do not round off to the nearest whole number and then simply add .0 to get extra significant digits. For example, if asked how many moles are found in 2.49 g of Ni, the molar mass for Ni from the periodic chart should be 58.69 g/mole or 58.7 g/mole, NOT 59 g/mol nor 59.0 g/mol.

1. Calculate the molar mass for the following compounds with the aid of the short list of atomic masses.

				Atomic Masses (g/mol)
(1)	$CHCl_3$	_____	Na	22.99
(2)	$CaCl_2$	_____	C	12.01
(3)	NH_4Cl	_____	O	16.00
(4)	$NaClO$	_____	Ca	40.08
(5)	Na_2CO_3	_____	Cl	35.45
			N	14.01
			H	1.008

Do your calculations here.

2. Using the molar masses that were calculated for the previous question, find the number of moles present for the following masses of chemical. Double-check your answer by making sure the units cancel out properly and by simple math logic to see if it makes sense (is the mass given more or less than the mass of 1 mole).

(1) 77 g $CHCl_3$
Number of moles = _____

(2) 31 g of $CaCl_2$
Number of moles = _____

(3) 89 g NH_4Cl
Number of moles = _____

Do your calculations here.

3.   Using the molar masses that were calculated for the first question, find the number of grams represented by the following number of moles of chemical. Again, double-check your answer logically.

(1)   1.5 moles of NaClO _____ g

(2)   0.67 mole of Na_2CO_3 _____ g

Do your calculations here.

4.   Using the partial list of atomic masses, calculate the number of moles represented by a certain mass of an element.

Na   22.99 g/mol
Ag   107.9 g/mol
S   32.07 g/mol

(1)   How many moles are represented by 0.460 g of Na atoms?

Do your calculations here.

(2)   How many moles are represented by 18.9 g of Ag?

Do your calculations here.

(3)   How many moles are represented by 79.5 g of S?

Do your calculations here.

5.   Using the following list of molar masses, calculate the elemental percent composition for the four compounds listed.

Molar masses   N = 14.01 g/mol,   O = 16.00 g/mol,   Al = 26.98 g/mol
C = 12.01 g/mol,   Mn = 54.94 g/mol,   K = 39.10 g/mol

(1) N_2O (2) Al_2O_3 (3) K_2CO_3 (4) $KMnO_4$

Do your calculations here.

(1) N_2O

(2) Al_2O_3

(3) K_2CO_3

(4) $KMnO_4$

6.   Fill in the blank space or circle the appropriate response.

The simplest formula of a compound, or the (1) _____ formula, tells us the (2) smallest/largest ratio of atoms present in a compound. The ratio is usually a small (3) _____ number ratio. It is possible for two compounds to have the same empirical formula, but different (4) _____ formulas. The true formula for a compound, or (5) _____ formula, represents the actual number of (6) _____ of each element found in one molecule of a compound. The mass of all the (7) _____ in a molecular formula is the compound's (8) _____ _____.

7. From the percent composition data given for each compound, calculate the empirical formula.

    (1) 86.6% Pb   Molar mass Pb = 207.2 g/mol
         13.4% S   Molar mass S = 32.07 g/mol
                                     Empirical formula _____

    Do your calculations here.

    (2) 69.59% Ba   Molar mass Ba = 137.3 g/mol
         6.08% C   Molar mass C = 12.01 g/mol
         24.33% O   Molar mass O = 16.00 g/mol
                                     Empirical formula _____

    Do your calculations here.

8. A certain compound is found to contain 1.45 g of Na, 2.05 g of S, and 1.5 g of O. With the aid of the list of molar masses, calculate the empirical formula.

         Na   Molar mass   22.99 g/mol
         S   Molar mass   32.07 g/mol
         O   Molar mass   16.00 g/mol
                                     Formula _____

    Do your calculations here.

9. The molar mass of an unknown sugar substance is experimentally found to be 180 g/mol. The percent composition data are determined to be 40% carbon, 7.0% hydrogen, and 53% oxygen. Calculate first the empirical formula and then the molecular formula.

C   Molar mass   12.01 g/mol
H   Molar mass   1.008 g/mol
O   Molar mass   16.00 g/mol
                                        Empirical _____

                                        Molecular _____

Do your calculations here.

10. A compound of bromine and iodine is formed by direct reaction between the two elements. It is found that 5.40 g of bromine react with 8.58 g of iodine. What is the empirical formula for the compound and if the molar mass is 206.8 g/mol, what is the molecular formula? What is the percent composition of the compound?

Do your calculations here.

11. Calculate the number of grams and the number of atoms represented by the following quantities of two elements.

(1)   2.25 mol of Ni   (2)   0.036 mol of Ag
        Molar mass of Ni $=$ 58.69 g/mol   Molar mass of Ag $=$ 107.9 g/mol

Do your calculations here.

Challenge Problems

12.   A compound known as cadaverine (1,5-pentane diamine) is a ptomaine formed by the action of bacteria on meat and fish. Analysis shows that the elemental composition is C 58.8%, H 13.8% and N 27.4%. Determine the empirical formula and the molecular formula. ("Pentane" means 5 carbon atoms.)

   Do your calculations here.

13.   How many grams of Fe contain the same number of atoms as 154 g of arsenic?

   Do your calculations here.

14.   At the present time the population of the United States is approximately 298,000,000. If one mole of pennies were distributed among the entire population, how many dollars would each person receive?

   Do your calculations here.

Chapter 7 puts you to work using the periodic table and chemical formulas to try out your math skills on some basic chemical problem solving. You will be interested to know that, prior to the late 1890s, many chemists spent their careers working on problems of the elemental composition of various compounds. The techniques and mathematical steps that they used are all related to the problems that you have just solved. Even today, the simple mathematical steps encountered in this chapter are routinely used by chemists just about every time they carry out a chemical reaction. These calculations will be built upon in later chapters so it is essential to grasp these fundamentals now.

ANSWERS TO QUESTIONS AND SOLUTIONS TO PROBLEMS

1.   (1) 12.01 g/mol + 1.008 g/mol + (3 x 35.45 g/mol) = 119.37 g/mol

       (2) 40.08 g/mol + (2 x 35.45 g/mol) = 110.98 g/mol

       (3) 14.01 g/mol + (4 x 1.008 g/mol) + 35.45 g/mol = 53.49 g/mol

       (4) 22.99 g/mol + 35.45 g/mol +16.00 g/mol = 74.44 g/mol

       (5) (2 x 22.99 g/mol) + 12.01 g/mol + (3 x 16.00 g/mol) = 105.99 g/mol

2.   (1)   molar mass of $CHCl_3$ = 119.37 g

       Therefore,

$$(77\ g)\left(\frac{1\ mol}{119.37\ g}\right) = 0.65\ mol$$

      (2)   molar mass of $CaCl_2$ = 110.98 g

       Therefore,

$$(31\ g)\left(\frac{1\ mol}{110.98\ g}\right) = 0.28\ mol$$

      (3)   molar mass of NH_4Cl = 53.49 g

       Therefore,

$$(89\ g)\left(\frac{1\ mol}{53.49\ g}\right) = 1.7\ mol$$

3.   (1)   1.5 moles of NaClO will equal

$$(1.5 \; \cancel{mol})\left(74.44 \; \frac{g}{\cancel{mol}}\right) = 1.1 \times 10^2 \, g$$

     (2)   0.67 mole of Na_2CO_3 will equal

$$(0.67 \; \cancel{mol})\left(\frac{105.99 \; g}{1 \; \cancel{mol}}\right) = 71 \, g$$

4.   (1) 0.0200 moles of Na

You are asked how many moles are contained in 0.460 g of Na atoms. One mole of any element is equal to the atomic mass expressed in grams. For Na this amount is 22.99 g. What we have said is that 22.99 g of Na equals 1.0 mol. We have 0.460 g, which is less than 1.0 mole. Arranging the problem so that the units will cancel out properly and give us an answer that is less than one, we have

$$(0.460 \; \cancel{g})\left(\frac{1 \; mol}{22.99 \; \cancel{g}}\right) = 0.0200 \, mol$$

     (2)   0.175 mole

1 mole of Ag = 107.9 g

Therefore,

$$(18.9 \; \cancel{g})\left(\frac{1 \; mol}{107.9 \; \cancel{g}}\right) = 0.175 \, mol$$

     (3)   2.48 moles of S

The problem is solved just as the others have been.

$$(79.5 \; \cancel{g})\left(\frac{1 \; mol}{32.07 \; \cancel{g}}\right) = 2.48 \, mol \; S$$

5.   (1) N_2O

molar mass $= (2 \times 14.01 \; g/mol) + 16.00 \; g/mol = 44.02 \; \dfrac{g}{mol}$

Therefore,

$$\%N = \frac{2 \times 14.01 \; \cancel{g/mol}}{44.02 \; \cancel{g/mol}} \times 100 = 63.65\%$$

$$\%O = \frac{16.00 \; \cancel{g/mol}}{44.02 \; \cancel{g/mol}} \times 100 = 36.35\%$$

(2) Al_2O_3

molar mass = $(2 \times 26.98 \text{ g/mol}) + (3 \times 16.00 \text{ g/mol}) = 102.0 \dfrac{\text{g}}{\text{mol}}$

Therefore,

$$\%Al = \frac{2 \times 26.98 \text{ g/mol}}{102.0 \text{ g/mol}} \times 100 = 52.90\%$$

$$\%O = \frac{3 \times 16.00 \text{ g/mol}}{102.0 \text{ g/mol}} \times 100 = 47.06\%$$

(3) K_2CO_3

molar mass = $(2 \times 39.10 \text{ g/mol}) + 12.01 \text{ g/mol} + (3 \times 16.00 \text{ g/mol}) = 138.2 \dfrac{\text{g}}{\text{mol}}$

Therefore,

$$\%K = \frac{2 \times 39.10 \text{ g/mol}}{138.2 \text{ g/mol}} \times 100 = 56.58\%$$

$$\%C = \frac{12.01 \text{ g/mol}}{138.2 \text{ g/mol}} \times 100 = 8.690\%$$

$$\%O = \frac{3 \times 16.00 \text{ g/mol}}{138.2 \text{ g/mol}} \times 100 = 34.73\%$$

(4) $KMnO_4$

molar mass = $39.10 \text{ g/mol} + 54.94 \text{ g/mol} + (4 \times 16.00 \text{ g/mol}) = 158.0 \dfrac{\text{g}}{\text{mol}}$

Therefore,

$$\%K = \frac{39.10 \text{ g}}{158.0 \text{ g}} \times 100 = 24.75\%$$

$$\%Mn = \frac{54.94 \text{ g}}{158.0 \text{ g}} \times 100 = 34.77\%$$

$$\%O = \frac{4 \times 16.00 \text{ g}}{158.0 \text{ g}} \times 100 = 40.51\%$$

6. (1) empirical   (2) smallest   (3) whole   (4) molecular
   (5) molecular   (6) atoms   (7) atoms   (8) molar mass

7. (1) PbS   (2) $BaCO_3$

(1) Taking 100 g of the compound composed of Pb and S, we would have 87 g Pb and 13 g S. We need to determine the number of moles of Pb and S present in the 100 g.

Therefore,

$$\text{Pb} \quad (87 \text{ g})\left(\frac{1 \text{ mol}}{207.2 \text{ g}}\right) = 0.42 \text{ mol}$$

$$\text{S} \quad (13 \text{ g})\left(\frac{1 \text{ mol}}{32.07 \text{ g}}\right) = 0.41 \text{ mol}$$

The ratio of Pb to S is 0.42 to 0.41 or 1:1. The formula is PbS.

(2)   Likewise, in a 100 g sample of the Ba, C, O compound, we would have 69.59 g of Ba, 6.08 g of C, and 24.33 g of O. The number of moles of each element would be:

$$\text{Ba} \quad (69.59 \text{ g})\left(\frac{1 \text{ mol}}{137.3 \text{ g}}\right) = 0.5068 \text{ mol}$$

$$\text{C} \quad (6.08 \text{ g})\left(\frac{1 \text{ mol}}{12.01 \text{ g}}\right) = 0.506 \text{ mol}$$

$$\text{O} \quad (24.33 \text{ g})\left(\frac{1 \text{ mol}}{16.00 \text{ g}}\right) = 1.521 \text{ mol}$$

To eliminate the decimals from our ratio we must divide each of the numbers by the smallest number.

$$\text{Ba} = \frac{0.5068}{0.506} = 1 \qquad\qquad \text{O} = \frac{1.521}{0.506} = 3$$

$$\text{C} = \frac{0.506}{0.506} = 1$$

When the ratio is expressed as small whole numbers, the correct empirical formula becomes $BaCO_3$.

8. $Na_2S_2O_3$

We must determine the number of moles of each element present and then find the smallest whole-number ratio.

		Smallest Ratio
Na $\quad (1.45 \text{ g})\left(\dfrac{1 \text{ mol}}{22.99 \text{ g}}\right) = 0.0631 \text{ mole}$		$\dfrac{0.0631}{0.0631} = 1$
S $\quad (2.05 \text{ g})\left(\dfrac{1 \text{ mol}}{32.07 \text{ g}}\right) = 0.0639 \text{ mole}$		$\dfrac{0.0639}{0.0631} = 1$
O $\quad (1.5 \text{ g})\left(\dfrac{1 \text{ mol}}{16.00 \text{ g}}\right) = 0.094 \text{ mole}$		$\dfrac{0.094}{0.0631} = 1.5$

Therefore, $1 : 1 : 1.5 = 2 : 2 : 3$

The empirical formula is $Na_2S_2O_3$

9. Empirical formula: CH_2O Molecular formula: $C_6H_{12}O_6$

The first step is to find the empirical formula from the number of moles of each element for a hypothetical 100 g of compound.

Smallest Ratio

$$C \quad (40.\,g)\left(\frac{1\,mol}{12.01\,g}\right) = 3.3\ moles \qquad\qquad \frac{3.3}{3.3} = 1$$

$$H \quad (7.0\,g)\left(\frac{1\,mol}{1.008\,g}\right) = 6.9\ moles \qquad\qquad \frac{6.9}{3.3} = 2.1$$

$$O \quad (53\,g)\left(\frac{1\,mol}{16.00\,g}\right) = 3.3\ moles \qquad\qquad \frac{3.3}{3.3} = 1$$

The rounded-off whole-number ratio would be $(CH_2O)_n$ and the empirical formula would therefore be CH_2O.

The molecular formula is calculated from the total atomic masses in the empirical formula and the molar mass. The total mass of the empirical formula is $12.01 + 2.016 + 16.00 = 30.03$. Next, determine the ratio between the given molar mass and the empirical formula mass.

$$\frac{180\ \cancel{g/mol}}{30.03\ \cancel{g/mol}} = 5.99$$

Therefore, we need to multiply the empirical formula by 6 to obtain the molecular formula.

$$6 \times CH_2O \text{ would be } C_6H_{12}O_6$$

10. Empirical formula and molecular formula are the same, BrI. The percent composition is 38.6% Br_2 and 61.3% I_2.

$$\text{Number of moles of } Br_2 = \frac{5.40\ \cancel{g}}{79.90\ \cancel{g}/mol} = 0.0676\ mol$$

$$\text{Number of moles of } I_2 = \frac{8.58\ \cancel{g}}{126.9\ \cancel{g}/mol} = 0.0676\ mol$$

The number of moles of each element are in a ratio of 1:1, so the empirical formula is BrI. The mass of one mole of BrI is 206.8 g, which matches the value given in the problem, so the molecular formula is also BrI. Percent composition is calculated as follows:

$$\% \, Br = \frac{mass\ Br_2}{total\ mass} = \frac{5.40\ \cancel{g}}{(5.40 + 8.58)\cancel{g}} \times 100 = 38.6\%$$

Likewise

$$\% \, I_2 = \frac{8.58\ \cancel{g}}{13.98\ g} \times 100 = 61.4\%$$

11.   (1) 132 g of Ni and 1.35×10^{23} atoms of Ni

Nickel has 58.69 g in 1 mole. We have 2.25 moles, which will be more than 58.69 g. Arranging the problem so that the units cancel out, we have:

$$(2.25 \ \cancel{mol})\left(58.69 \ \frac{g}{\cancel{mol}}\right) = 132 \text{ g of Ni}$$

To find the number of atoms in a certain number of moles, we simply multiply the number of moles by Avogadro's number.

$$(2.25 \ \cancel{mol})\left(6.022 \times 10^{23} \ \frac{\text{atoms}}{\cancel{mol}}\right) = 1.35 \times 10^{24} \text{ atoms Ni}$$

(2)   3.9 g of Ag and 2.2×10^{22} atoms Ag

$$(0.036 \ \cancel{mol})\left(107.9 \ \frac{g}{\cancel{mol}}\right) = 3.9 \text{ g of Ag}$$

$$(0.036 \ \cancel{mol})\left(6.022 \times 10^{23} \ \frac{\text{atoms}}{\cancel{mol}}\right) = 2.2 \times 10^{22} \text{ atoms Ca}$$

12.   The empirical and molecular formula are both $C_5H_{14}N_2$, since we know from the problem that the compound contains 5 C atoms.

$$C \quad \frac{58.5 \ \cancel{g}}{12.01 \ \cancel{g}/mol} = 4.90 \text{ mol}$$

$$N \quad \frac{27.4 \ \cancel{g}}{14.01 \ \cancel{g}/mol} = 1.96 \text{ mol}$$

$$H \quad \frac{13.8 \ \cancel{g}}{1.008 \ \cancel{g}/mol} = 13.7 \text{ mol}$$

13.   115 g Fe. We can start by finding out how many moles of arsenic are represented by 154 g.

$$\frac{154 \ \cancel{g}}{74.92 \ \cancel{g}/mol} = 2.06 \text{ mol As}$$

2.06 moles of As contains the same number of atoms as 2.06 moles of Fe. To convert this value into grams of Fe, we need only to multiply the molar mass by the number of moles.

$$(2.06 \ \cancel{mol})\left(55.85 \ \frac{\text{g Fe}}{\cancel{mol}}\right) = 115 \text{ g Fe (3 significant figures)}$$

14. $\$2.27 \times 10^{13}$ per person. One mole of pennies equals one Avogadro's number of pennies of 6.022×10^{23} pennies. To find dollars, divide this number by 10^2 or 100.

$$\frac{6.022 \times 10^{23} \ \cancel{\text{pennies}}}{1 \times 10^2 \ \frac{\cancel{\text{pennies}}}{\text{dollar}}} = 6.022 \times 10^{21} \text{ dollars}$$

$$\frac{6.022 \times 10^{21} \text{ dollars}}{2.65 \times 10^8 \text{ persons}} = \$2.27 \times 10^{13}/\text{person}$$

Chemical Equations

SELECTED CONCEPTS REVISITED

Understanding and balancing chemical equations is a key aspect of any chemistry course. This is how chemists communicate reactions to each other, regardless of language. When balancing an equation, you cannot change the molecular formula of any substance. Only the coefficients of the substances may change. Therefore be certain to establish the molecular formula for all the reactants and products <u>before</u> you begin to balance the reaction. You may only change the amount of each substance present by changing the coefficients of the compounds.

When you place a coefficient in front of a formula it multiples every atom in the formula by that amount. Therefore when you are counting atoms, multiply any subscripts by the coefficient to get the number of that type of atom present. It is also useful to remember that the subscript after a set of atoms in parentheses applies to each atom in the parentheses.

For example, $3\ Al_2(SO_4)_3$ has:
(3)(2) Al = 6 Al
(3)(1)(3) S = 9 S
(3)(4)(3) O = 36 O

In many cases where there are identical polyatomic ions on both sides of the equation, you can simply count the polyatomics as a unit and balance them instead of breaking them down into individual atoms. This shortcut is most useful when balancing double-displacement reactions.

Balancing chemical reactions is largely trial and error. You will only get proficient at balancing equations by practicing. However, this is one of the easiest types of problems to double-check to see if you have the right answer – either the atoms are balanced when you are finished or they are not. Always check your balanced equation once you think it is properly balanced.

You have also been introduced to the activity series for metals. Notice the general trend of the more reactive the metal is, the easier it is to replace H in water or acids. It is easier to replace H from an acid than from water and to replace an H from hot water than from cold water. So only the most reactive metals will react with cold water.

In a chemical equation, we write "+ heat", "not − heat". For example, an exothermic reaction would be written as

$$A \rightarrow B + heat \qquad NOT \qquad A - heat \rightarrow B$$

COMMON PITFALLS TO AVOID

Do not try to do more than one step at a time when balancing equations. Make sure you have the right formula for the reactants and products and then start to balance your equation changing only the coefficients. It may help you to think of it as two problems – (1) write the molecular formula for each reactant and product (this involves putting the right subscripts on the atoms involved) and (2) balance the equation (this involves changing the coefficients).

When you write the molecular formula for the products of a double displacement reaction do not forget you must balance the charges to determine the formula. You will be putting together a cation and an anion but the subscripts of the cation and the anion (i.e. the subscripts in the formula) are not necessarily the same as the subscripts associated with them in the reactants. Once you know the formula and charge of each ion, ignore the reactants and determine the formula for the compound based on the relative charges of the ions. If you are comfortable with naming ionic compounds you can often double-check the formula of double-displacement reaction by determining the name of the compound first and then writing the formula.

For example, what are the formulas for the products when $AgNO_3$ reacts with $MgCl_2$?
Here is one possible approach: you know the cation from the first compound will combine with the anion from the second compound so name the respective ions (silver ion and chloride ion) so one product is silver chloride. Then simply write the formula for silver chloride, $AgCl$ because you are combining Ag^+ and Cl^- (not $AgCl_2$ which incorrectly assumes that the subscript associated with the ion in the reactant is the same as the subscript in the product).

SELF-EVALUATION SECTION

1.   Match the symbols used in chemical equations with the corresponding descriptive statements.

Symbols

$\rightarrow$	(s)
$+$	(l)
$\rightleftarrows$	(g)
	(Δ)
	(aq)

(1)   Gas (written after substance) _____
(2)   Reversible reaction; equilibrium between reactants and products _____
(3)   Heat _____
(4)   Added to _____
(5)   Liquid (written after substance) _____
(6)   Aqueous solution (substance dissolved in water) _____
(7)   Yields; produces (points to products) _____
(8)   Solid (written after substance) _____

2. Consider the equation $2 Al(OH)_3 + 3 H_2SO_4 \rightarrow Al_2(SO_4)_3 + 6 H_2O$

Label the following parts of the equation: subscripts, reactants, coefficients, products.

3. Balance the following equations or translate the word equations in formulas and then balance them. The first two word-to-formula equations emphasize the process involved. You should apply the same procedure whenever you are asked to write a balanced formula equation.

(1) $Fe + H_2O \rightarrow Fe_3O_4 + H_2$

(2)   Potassium nitrate $\rightarrow$ Potassium nitrite + Oxygen
       formula: _____ $\rightarrow$ _____   + _____
       balanced equation:

(3)   Calcium oxide + Hydrochloric acid $\rightarrow$ Calcium chloride + Water
       formula: _____ + _____ $\rightarrow$ _____   + _____
       balanced equation:

(4) $H_2O_2 \rightarrow H_2O + O_2$

(5) $NH_4NO_2 \rightarrow N_2 + H_2O$

(6)   Copper metal + Sulfuric acid $\rightarrow$ Copper (II) sulfate + Water + Sulfur dioxide

(7)   Bromine + Hydrogen sulfide $\rightarrow$ Hydrogen bromide + Sulfur

(8) $C_6H_{14} + O_2 \rightarrow CO_2 + H_2O$

(9) $CO + Fe_3O_4 \rightarrow FeO + CO_2$

(10) Zinc sulfide + Oxygen → Zinc oxide + Sulfur dioxide

(11) $K + H_2O \rightarrow KOH + H_2$

(12) Carbon + Oxygen → Carbon monoxide

(13) $Ca + O_2 \rightarrow CaO$

(14) Sodium hydrogen carbonate + Sulfuric acid → Sodium sulfate + Water + Carbon dioxide

(15) $N_2 + H_2 \rightarrow NH_3$

4. Identify the following reactions as combinations (C), decomposition (D), single displacement (SD), or double displacement (DD).

(1) $2 \, Al(OH)_3 + 3 \, H_2SO_4 \rightarrow Al_2(SO_4)_3 + 6 \, H_2O$ _____

(2) $4 \, K + O_2 \rightarrow 2 \, K_2O$ _____

(3) $Cl_2 + 2 \, NaBr \rightarrow Br_2 + 2 \, NaCl$ _____

(4) $2 \, HgO \rightarrow 2 \, Hg + O_2$ _____

(5) $MgCl_2 + 2 \, AgNO_3 \rightarrow 2 \, AgCl + Mg(NO_3)_2$ _____

(6) $CaO + H_2O \rightarrow Ca(OH)_2$ _____

(7) $2 \, HCl + Na_2CO_3 \rightarrow 2 \, NaCl + H_2O + CO_2$ _____

(8) $2 \, KClO_3 \rightarrow 2 \, KCl + 3 \, O_2$ _____

5. Identify the following reactions as exothermic (exo) or endothermic (endo).

(1) $N_2(g) + O_2(g) + 181 \, kJ \rightarrow 2 \, NO(g)$ _____

(2) $C(s) + O_2(g) \rightarrow CO_2(g) + 94.0 \, kcal$ _____

(3) $C_3H_8(g) + 5 \, O_2(g) \rightarrow 3 \, CO_2(g) + 4 \, H_2O(g) + 2200 \, kJ$ _____

6. Interpret the odd-numbered reactions of question 4 in terms of number of moles of reactants and products involved. Write your answers below.

(1)

(3)

(5)

(7)

Challenge Problems

7. Identify each of following reactions as combination, decomposition, single displacement or double displacement. Complete and balance each one, converting names into formula when necessary.

(1) potassium chromate + lead (II) nitrate $\longrightarrow$

(2) $Na + H_2O \longrightarrow$

(3) $HCl + K_2CO_3 \longrightarrow$

(4) $H_2 + N_2 \xrightarrow{\Delta}$

(5) $CaCO_3 \xrightarrow{\Delta}$

(6) $Zn + Pb(NO_3)_2 \longrightarrow$

(7) $Al_2(SO_4)_3 + NH_4OH \longrightarrow$

(8)   Calcium oxide + water $\longrightarrow$

(9) $Cl_2 + KBr \rightarrow$

(10) $Mg + NiCl_2 \longrightarrow$

(11) $Ba(NO_3)_2 + Na_2SO_4 \longrightarrow$

(12) $NH_3 + HCl \longrightarrow$

(13) $H_2 + Cl_2 \longrightarrow$

(14) nitric acid + sodium hydroxide $\longrightarrow$

(15) $NaCl + H_2SO_4 \longrightarrow$

(16) $Mg + H_2SO_4 \longrightarrow$

(17) $Na_2O + H_2SO_4 \longrightarrow$

(18) $Mg + O_2 \xrightarrow{\Delta}$

(19) $NaI + Cl_2 \longrightarrow$

(20) silver nitrate + sodium chloride $\longrightarrow$

8.   Will a reaction occur when the following are mixed? If so, write a balanced equation. Use the short activity series in the chapter.

(1) $Al(s) + NaBr(aq)$ (4) $Zn(s) + CuCl_2(aq)$ (7) $Fe(s) + H_2O(l)$

(2) $Cu(s) + HCl(aq)$ (5) $Mg(s) + NiCl_2(aq)$ (8) $Ni(s) + H_2O(g)$

(3) $K(s) + H_2O$ (6) $Sn(s) + H_2O(g)$

9.   An aqueous reaction with this energy profile was carried out in a beaker. If you touched the beaker, would the glass be hot or cold? Why? Would heat be written as a reactant or product for this reaction?

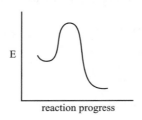

E

reaction progress

10.

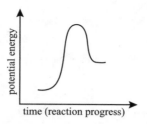

potential energy

time (reaction progress)

(1)   Does the potential energy diagram show an exothermic or an endothermic reaction?

(2)   Would the energy term be written as a reactant or product in the chemical equation?

(3)   On the graph, show clearly the activation energy for the reaction.

RECAP SECTION

Chapter 8 is a self-contained section similar to Chapter 6. You should refer to both of these chapters from time to time to review nomenclature and chemical equations. You have learned what a chemical equation is and how to put one together in a balanced form. We are now able to use a chemical equation rather than a word equation to describe chemical changes.

ANSWER TO QUESTIONS

1.   (1) (g) (2) $\leftrightarrows$ (3) Δ (4) $+$
      (5) (l) (6) (aq) (7) $\rightarrow$ (8) (s)

2. $2\,Al(OH)_3$ $+$ $3\,H_2SO_4$ $\rightarrow$ $Al_2(SO_4)_3$ $+$ $6\,H_2O$
      Coefficients   2   3   1   6
      Reactants $Al(OH)_3$ H_2SO_4
      Subscripts (shown in bold) $2\,Al(OH)_3$ $+$ $3\,H_2SO_4$ $\rightarrow$ $Al_2(SO_4)_3$ $+$ $6\,H_2O$
      Products $Al_2(SO_4)_3$ H_2O

3.   (1) $3\,Fe\,+\,4\,H_2O\,\rightarrow\,Fe_3O_4\,+\,4\,H_2$
      (2)   formula: $KNO_3\,\rightarrow\,KNO_2\,+\,O_2$
             balanced equation: $2\,KNO_3\,\rightarrow\,2\,KNO_2\,+\,O_2$
      (3)   formula: $CaO\,+\,HCl\,\rightarrow\,CaCl_2\,+\,H_2O$
             balanced equation: $CaO\,+\,HCl\,\rightarrow\,CaCl_2\,+\,H_2O$
      (4) $2\,H_2O_2\,\rightarrow\,2\,H_2O\,+\,O_2$
      (5) $NH_4NO_2\,\rightarrow\,N_2\,+\,2\,H_2O$
      (6) $Cu\,+\,2\,H_2SO_4\,\rightarrow\,CuSO_4\,+\,2\,H_2O\,+\,SO_2$
      (7) $Br_2\,+\,H_2S\,\rightarrow\,2\,HBr\,+\,S$
      (8) $2\,C_6H_{14}\,+\,19\,O_2\,\rightarrow\,12\,CO_2\,+\,14\,H_2O$
      (9) $CO\,+\,Fe_3O_4\,\rightarrow\,3\,FeO\,+\,CO_2$
      (10) $2\,ZnS\,+\,3\,O_2\,\rightarrow\,2\,ZnO\,+\,2\,SO_2$

(11) $2 K + 2 H_2O \rightarrow 2 KOH + H_2$

(12) $2 C + O_2 \rightarrow 2 CO$

(13) $2 Ca + O_2 \rightarrow 2 CaO$

(14) $2 NaHCO_3 + H_2SO_4 \rightarrow Na_2SO_4 + 2 H_2O + 2 CO_2$

(15) $N_2 + 3 H_2 \rightarrow 2 NH_3$

4.   (1) DD   (2) C   (3) SD   (4) D
     (5) DD   (6) C   (7) DD   (8) D

5.   (1) endo   (2) exo   (3) exo

6.   (1) Reactants 2 moles $Al(OH)_3$ and 3 moles H_2SO_4
          Products   1 mole $Al_2(SO_4)_3$ and 6 moles H_2O

     (3) Reactants 1 mole Cl_2 and 2 moles $NaBr$
          Products   1 mole Br_2 and 2 moles $NaCl$

     (5) Reactants 1 mole $MgCl_2$ and 2 moles $AgNO_3$
          Products   2 moles $AgCl$ and 1 mole $Mg(NO_3)_2$

     (7) Reactants 2 moles HCl and 1 mole Na_2CO_3
          Products   2 moles $NaCl$ and 1 mole H_2O and 1 mole CO_2

7.   (1) double displacement $K_2CrO_4 + Pb(NO_3)_2 \rightarrow 2KNO_3 + PbCrO_4$
     (2) single displacement $2 Na + 2 H_2O \rightarrow 2 NaOH + H_2(g)$
     (3) double displacement $2 HCl + K_2CO_3 \rightarrow 2 KCl + H_2O + CO_2(g)$
     (4) combination $3 H_2 + N_2 \overset{\Delta}{\rightarrow} 2 NH_3$
     (5) decomposition $CaCO_3 \overset{\Delta}{\rightarrow} CaO + CO_2(g)$
     (6) single displacement $Zn + Pb(NO_3)_2 \rightarrow Zn(NO_3)_2 + Pb$
     (7) double displacement $Al_2(SO_4)_3 + 6 NH_4OH \rightarrow 3 (NH_4)_2SO_4 + 2 Al(OH)_3$
     (8) combination $CaO + H_2O \rightarrow Ca(OH)_2$
     (9) single displacement $Cl_2 + 2 KBr \rightarrow Br_2 + 2 KCl$
     (10) single displacement $Mg + NiCl_2 \rightarrow Ni + MgCl_2$
     (11) double displacement $Ba(NO_3)_2 + Na_2SO_4 \rightarrow BaSO_4 + 2 NaNO_3$
     (12) combination $NH_3 + HCl \rightarrow NH_4Cl$
     (13) combination $H_2 + Cl_2 \rightarrow 2 HCl$
     (14) double displacement $HNO_3 + NaOH \rightarrow NaNO_3 + H_2O$
     (15) double displacement $NaCl + H_2SO_4 \rightarrow NaHSO_4 + HCl(g)$
     (16) single displacement $Mg + H_2SO_4 \rightarrow MgSO_4 + H_2(g)$
     (17) double displacement $Na_2O + H_2SO_4 \rightarrow Na_2SO_4 + H_2O$

(18)	combination	$2\,Mg + O_2 \overset{\Delta}{\rightarrow} 2\,MgO$
(19)	single displacement	$2\,NaI + Cl_2 \rightarrow 2\,NaCl + I_2$
(20)	double displacement	$AgNO_3 + NaCl \rightarrow AgCl + NaNO_3$

8.   (1)   No

       (2)   No

       (3)   Yes $2\,K(s) + 2\,H_2O \rightarrow 2\,KOH + H_2(g)$

       (4)   Yes $Zn(s) + CuCl_2(aq) \rightarrow Cu(s) + ZnCl_2(aq)$

       (5)   Yes $Mg(s) + NiCl_2(aq) \rightarrow Ni(s) + MgCl_2(aq)$

       (6)   No

       (7)   No (if H_2O were hot then it would have reacted)

       (8)   No

9.   The reaction profile shows an exothermic reaction that means heat is given off during the reaction and therefore the beaker should feel hotter after the reaction than it did before the reaction. In exothermic reactions heat can be written as a product.

10.   (1) endothermic   (2) reactant (net absorption of energy)

      (3)

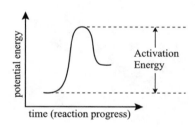

CHAPTER NINE

Calculations from Chemical Equations

SELECTED CONCEPTS REVISITED

This chapter deals with a lot of calculations and it is important that you begin to recognize the "power" of the mole. Two important relationships involving moles is molar mass and mole ratios. Molar mass allows us to relate a number to a mass, that is, it relates mass to moles as the name suggests. Therefore we can translate an easily measured quantity, the mass, to the mole which allows us to compare numbers of one substance to another. The mole ratio provides us with the numerical relationship between substances in a chemical reaction. If you understand these relationships and where they come from, working through the problems in this chapter will be much easier. The molar mass is available from the periodic table and the mole ratio is given in a balanced chemical equation.

When carrying out calculations it is crucial that you determine what quantity the question is asking for, what information is directly given, and what information is accessible from a periodic table. Remember that with a balanced chemical equation (for the mole ratio) and a periodic table (for the molar masses), you should be able to solve most of the problems in this chapter. Do not forget that Avogadro's number provides the relationship between actual numbers of particles and the mole.

The limiting reactant is the reactant that runs out first based on the stoichiometry of the reaction. We need to compare moles of one reactant to moles of another using the mole ratio. Whichever reactant is used up first will limit the amount of product that can be formed. The total amount of possible product that can be formed from the amounts of reactants given is called the theoretical yield. The theoretical yield assumes that 100% of the limiting reactant is converted to product. Since this rarely occurs in practice, we generally also calculate the percent yield of the reaction which compares the actual yield to the theoretical yield.

COMMON PITFALLS TO AVOID

You must have a balanced chemical equation in order to carry out most of the calculations or at the very least have some other way of determining the mole ratio. Do not forget to check if the equation given to you is balanced.

Use logic to avoid the tendency to use the mole ratio "upside down". Talk it out to yourself to see if it makes sense. For example, if 2 moles of product are produced for every 1 mole of reactant, then the moles of product should be greater than the moles of reactant. So if you start out with 0.68 moles of reactant, the moles of product must be greater than 0.68 so you must multiply by 2 (not divide). Dimensional analysis helps here also as long you make sure "moles reactant" and "moles product" are used instead of just the numerical ratio.

Check with your instructor on their policy of rounding off numbers after every step of a multi-step calculation. In general, since most of these calculations involve only multiplication/division one or two extra significant figures are carried through to the end of the calculation before the final rounding off to avoid rounding error.

The coefficients of a balanced chemical equation are irrelevant when dealing with molar mass. By definition the molar mass is the mass of one mole of a substance. The number of moles of the substance being used is addressed in the mole ratio not in the molar mass. Do not use the coefficients for the molar mass.

Understand what the question is asking you to find. Too many students memorize the process: g of A to moles of A to moles of B to g of B. Then, given moles of A and asked to find g of B, students often incorrectly start the calculation using the molar mass of A (which is not needed here). The question already assumed you knew the moles of A. Read and understand the questions and understand the relationships (molar mass and mole ratio) and what you can calculate using each relationship.

SELF-EVALUATION SECTION

1. Fill in the blanks with the appropriate numbers that reflect the necessary moles to keep the equation balanced.

$$2\,Na + Cl_2 \rightarrow 2\,NaCl$$

   (1) 2 moles of Na reacts with _____ moles of Cl_2 to produce _____ moles of NaCl.

   (2) 0.50 moles of Na reacts with _____ moles of Cl_2 to produce _____ moles of NaCl.

   (3) _____ moles of Na reacts with 0.040 moles of Cl_2 to produce _____ moles of NaCl.

   (4) _____ moles of Na reacts with _____ moles of Cl_2 to produce 1.2 moles of NaCl.

$$Pt + 8\,HCl + 2\,HNO_3 \rightarrow H_2PtCl_6 + 2\,NOCl + 4\,H_2O$$

   (5) 0.300 _____ _____ _____ _____ _____ moles

   (6) _____ _____ 0.016 _____ _____ _____ moles

   (7) _____ _____ _____ _____ _____ 0.100 moles

2. For the combustion of hexane as written in the equation

$$2\,C_6H_{14} + 19\,O_2 \rightarrow 12\,CO_2 + 14\,H_2O$$

   (1) what is the mole ratio of C_6H_{14} to CO_2? _____

   (2) what is the mole ratio of C_6H_{14} to H_2O? _____

   (3) Given 4 moles of C_6H_{14} and 40 moles of O_2 at the start of the reaction, which is the limiting reagent?

   _____

   (4) How many moles of CO_2 can be produced from 5 moles of C_6H_{14}?

(5)   If only 3.0 moles of C_6H_{14} is available for the reaction, what is the theoretical yield of H_2O in moles? _____ in grams? _____ of CO_2 in grams? _____

(6)   After carrying out the above reaction with 3.0 moles of C_6H_{14}, a chemist measured an actual yield of 550 g of CO_2. What was the percent yield of CO_2?

Do your calculations here.

3.   For each of the following reactions, choose the sequence of steps needed. This will help you to recognize the steps needed depending on the information given in the question.

(1) $2\,ZnS\ +\ 3\,O_2\ \rightarrow\ 2\,ZnO\ +\ 2\,SO_2$

A chemist weighed 100. g of ZnS and burned it under the appropriate conditions. What is the theoretical yield in moles of sulfur dioxide?

(a) g ZnS $\rightarrow$ moles ZnS $\rightarrow$ g SO_2 $\rightarrow$ moles SO_2
(b) g ZnS $\rightarrow$ moles ZnS $\rightarrow$ moles SO_2
(c) g ZnS $\rightarrow$ g SO_2 $\rightarrow$ mol SO_2
(d) g ZnS $\rightarrow$ g O_2 $\rightarrow$ mol O_2 $\rightarrow$ mol SO_2

(2) $Na_2O\ +\ H_2O\ \rightarrow\ 2\,NaOH$

If an excess of water was mixed with 25.0 g of Na2O under reaction conditions, what is the theoretical mass of the product?

(a) g Na_2O x 2 $\rightarrow$ g NaOH
(b) g Na_2O $\rightarrow$ mol NaOH $\rightarrow$ g NaOH
(c) g Na_2O $\rightarrow$ mol Na_2O $\rightarrow$ g NaOH
(d) g Na_2O $\rightarrow$ mol Na_2O $\rightarrow$ mol NaOH $\rightarrow$ g NaOH

4.   For the reaction in part (1) of the previous question, calculate the theoretical moles of ZnO obtained if 100. g of <u>each</u> reactant was used. (Hint, first determine the limiting reagent.)

Do your calculations here.

5.   We will now use the techniques of Chapter 9 to gain further chemical equation problem-solving skill.

Balance the following equation and then calculate the requested quantities.

$$PbO_2 \xrightarrow{\Delta} PbO + O_2(g)$$

(1)   How many grams of O_2 can be obtained from 100 grams of PbO_2? This is the theoretical yield. Remember that we need to (a) use a balanced equation, (b) determine the number of moles of starting substance, (c) calculate the number of moles of desired substance, and (d) convert moles to grams of desired substance.

(2)   What is the percent yield if the actual yield of O_2 in the above reaction was 5.0 grams?

Do your calculations here.

6.   Very often in industrial chemical processes, one of the reactants will be present in an amount that exceeds the requirements of the balanced equation. The reactant that is not in excess will therefore limit the amount of product that is formed and is named the limiting reactant. Using the equation given, answer the questions below.

$$2\ Al(OH)_3 + 3\ H_2SO_4 \rightarrow Al_2(SO_4)_3 + 6\ H_2O$$

(1)   The reaction is run with excess of sulfuric acid. What is the limiting reactant?

(2)   You have 9 moles of H_2SO_4 present for the reaction. How many moles of $Al(OH)_3$ are required to react completely with this amount of H_2SO_4?

(3)   If 4 moles of $Al(OH)_3$ is the amount available, how much of the 9 moles of H_2SO_4 can be used?

(4)   Using the 4 moles of $Al(OH)_3$, how many moles of $Al_2(SO_4)_3$ and H_2O will you be able to produce?

Do your calculations here.

7. In the following reaction, how many moles of ZnO can be obtained? Also determine which reactant is the limiting reactant and which reactant is in excess.

$$2\ ZnS + 3\ O_2 \rightarrow 2\ ZnO + 2\ SO_2$$
100. g   100. g

Do your calculations here.

8. Balance the following equation and calculate how many moles of Cu can be formed from 5.0 moles of Al and 10.0 moles of $CuSO_4$. What is the limiting reactant and how much of the excess reactant is left after the reaction?

$$Al + CuSO_4 \rightarrow Cu + Al_2(SO_4)_3$$
5.0 mol 10.0 mol

Do your calculations here.

9. What is the theoretical yield of Fe_2O_3 that can be produced from 3.0 kg of Fe according to the following unbalanced equation? Balance the equation first.

$$Fe + O_2 \rightarrow Fe_2O_3$$

Do your calculations here.

10. Consider the reaction $2\ MX_2 + X_2 \rightarrow 2\ MX_3$.

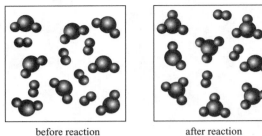 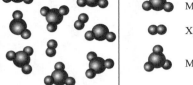

before reaction   after reaction

(1) Which reagent is the limiting reagent? (Hint, consider only the before picture.)
(2) If 0.342 moles of X_2 is used, what is the minimum amount of MX_2 needed for the reaction to go to completion?
(3) What is the actual yield of the reaction as illustrated?

Challenge Problems

11. It is possible to reclaim silver from used photographic fixer by using the active metal, powdered zinc. Zinc replaces the silver in solution, followed by conversion of the silver to silver oxide. After filtering, the silver oxide is reduced to metallic silver by carbon. The series of reactions are:

(1) $2\ AgBr + Zn \longrightarrow ZnBr_2 + 2\ Ag$

(2) $4\ Ag + O_2 \longrightarrow 2\ Ag_2O$

(3) $2\ Ag_2O + C \longrightarrow CO_2 + 4\ Ag$

How much silver can be reclaimed from 2.000 gallons of used fixer (density 1.018 g/mL) if the silver concentration is 300. parts per million? Part per million is a general purpose concentration term that can take on a variety of units. For example 1 μg per gram, 1 μL per liter and 1 mg per kilogram are examples of 1 ppm concentration.

Do your calculations here.

12. Refer to the problem above and the last reaction. The carbon that reduces the Ag_2O to metallic Ag comes from the filter paper that was used in the filtration step following reaction (2). Assuming the filter paper to be pure cellulose with an empirical formula of $C_6H_{12}O_6$, what mass of filter paper is required to reduce the 2.31 g of silver contained in the two gallon volume of fixer?

Do your calculations here.

The principal objective of Chapter 9 is to present a logical method for attacking problem solving associated with chemical equations. Since chemistry deals with chemical reactions and reactions are expressed in equation form, it follows that calculations associated with reactions are as relevant and practical as any topic in chemistry. Professionals in agriculture, home economics, forestry, biology, and chemical engineering, in addition to chemists, are constantly working with stoichiometric calculations.

The technique of solving mole stoichiometric problems is very important to learn. Once you have a balanced equation, it is possible to set up ratio between any two species in the equation. If the problem asks for grams of reactant to produce so many grams of product, there will be additional calculation steps to perform on each side of the ratio, but the ratio is the connecting link.

grams reactant A → moles reactant A → moles product B → grams product B

ANSWERS TO QUESTIONS AND SOLUTIONS TO PROBLEMS

1. (1) 1, 2   (2) 0.25, 0.50   (3) 0.080, 0.080   (4) 1.2, 0.60

(5) 0.300	2.40	0.600	0.300	0.600	1.20 moles
(6) 0.0080	0.13	0.016	0.0080	0.016	0.032 moles
(7) 0.0250	0.200	0.0500	0.0250	0.0500	0.100 moles

2. (1) 2:12   (2)1:7   (3) C_6H_{14}

(4) moles of CO_2 = 5 moles of $C_6H_{14} \times \dfrac{12 \text{ moles } CO_2}{2 \text{ moles } C_6H_{14}}$ = 30. moles CO_2

(5) moles of H_2O = 3.0 moles of $C_6H_{14} \times \dfrac{14 \text{ moles } H_2O}{2 \text{ moles } C_6H_{14}}$ = 21 moles H_2O

g of H_2O = 21 moles $H_2O \times \dfrac{18.0 \text{ g}}{\text{mol}}$ = 378 g H_2O = 380 g H_2O

g of CO_2 = 3.0 moles of $C_6H_{14} \times \dfrac{12 \text{ moles } CO_2}{2 \text{ moles } C_6H_{14}} \times \dfrac{44.0 \text{ g } CO_2}{\text{mol } CO_2}$ = 792 g = 790 g

(6) % yield = $\dfrac{550 \text{ g}}{790 \text{ g}} \times 100$ = 69.62 = 69% yield of CO_2

3. (1) (b)   (2) (d)

4. (1) 100. g of ZnS and 100. g of O_2 used so we first need to convert to moles of each and then compare the stoichiometry.

Moles of ZnS $= 100.$ g of ZnS $\times \dfrac{\text{mole}}{97.4 \text{ g}} = 1.03$ moles ZnS

Moles of $O_2 = 100.$ g of $O_2 \times \dfrac{\text{mole}}{32.0 \text{ g}} = 3.13$ moles O_2

We need 2 moles of ZnS for every 3 moles of O_2; for 1.03 mols of ZnS, at least

1.03 moles ZnS $\times \dfrac{3 \text{ moles } O_2}{2 \text{ moles ZnS}} = 1.55$ moles O_2 would be needed.

Since 1.55 moles of O_2 would be needed and there is 3.13 moles of O_2 available, ZnS is the limiting reagent.

Theoretical moles of ZnO $= 1.03$ moles ZnS $\times \dfrac{2 \text{ moles ZnO}}{2 \text{ moles ZnS}} = 1.03$ moles ZnO

5. (1) 6.69 g of O_2
The equation must be balanced before we can determine a proper mole ratio for calculating the amount of O_2.

$$2 \text{ PbO}_2 \xrightarrow{\Delta} 2 \text{ PbO} + O_2(g)$$

Next, we need to convert 100. grams of PbO_2 into the number of moles of PbO_2.

The molar mass of $PbO_2 = 207.2$ g/mol $+ (2 \times 16.00$ g/mol$) = 239.2 \dfrac{\text{g}}{\text{mol}}$

Number of moles of $PbO_2 = (100. \text{ g})\left(\dfrac{1 \text{ mol}}{239.2 \text{ g}}\right) = 0.418$ mol

The mole ratio is $\dfrac{\text{mol desired substance}}{\text{mol starting substance}} = \dfrac{1 \text{ mol } O_2}{2 \text{ mol PbO}_2}$

Moles of $O_2 = (0.418 \text{ mol PbO}_2)\left(\dfrac{1 \text{ mol } O_2}{2 \text{ mol PbO}_2}\right) = 0.209$ mol O_2

To convert moles of O_2 into grams, we multiply the number of moles by the molar mass.

0.209 mol $\times 32.00 \dfrac{\text{g}}{\text{mol}} = 6.69$ g O_2

(2) 75%
The percent yield is determined by dividing the actual yield by the theoretical yield multiplied by 100.

$$\dfrac{\text{actual}}{\text{theoretical}} \times 100$$

$$\dfrac{5.0 \text{ g}}{6.7 \text{ g}} \times 100 = 75\%$$

6. (1) $Al(OH)_3$, aluminum hydroxide   (2) 6 moles   (3) 6 moles
    (4) 2 moles $Al_2(SO_4)_3$, 12 moles H_2O

7. 1.03 mole ZnO, ZnS is limiting reactant, O_2 is in excess. First we need to determine the number of moles of ZnO that can be obtained from each of the reactants.

$$(100 \text{ g ZnS})\left(\frac{1 \text{ mol ZnS}}{97.46 \text{ g ZnS}}\right)\left(\frac{2 \text{ mol ZnO}}{2 \text{ mol ZnS}}\right) = 1.03 \text{ mol ZnO}$$

$$(100 \text{ g } O_2)\left(\frac{1 \text{ mol } O_2}{32.00 \text{ g } O_2}\right)\left(\frac{2 \text{ mol ZnO}}{2 \text{ mol } O_2}\right) = 2.08 \text{ mol ZnO}$$

8. The balanced equation is

$$2 \text{ Al} + 3 \text{ CuSO}_4 \rightarrow 3 \text{ Cu} + \text{Al}_2(\text{SO}_4)_3$$

First we need to determine the number of moles of Cu that can be formed from each reactant.

$$(5.0 \text{ mol Al})\left(\frac{3 \text{ mol Cu}}{2 \text{ mol Al}}\right) = 7.5 \text{ mol Cu}$$

$$(10.0 \text{ mol CuSO}_4)\left(\frac{3 \text{ mol Cu}}{3 \text{ mol CuSO}_4}\right) = 10.0 \text{ mol Cu}$$

Therefore, Al is the limiting reactant and 7.5 mol of Cu can be formed. Next we need to calculate the number of moles of $CuSO_4$ that will react with 5.0 moles of Al.

$$(5.0 \text{ mol Al})\left(\frac{3 \text{ mol CuSO}_4}{2 \text{ mol Al}}\right) = 7.5 \text{ mol CuSO}_4$$

Therefore, 10.0 mol $CuSO_4$ − 7.5 mol $CuSO_4$ = 2.5 mol of $CuSO_4$ in excess.

9. 4300 g Fe_2O_3 (2 significant figures)
    The balanced equations is

$$4 \text{ Fe} + 3 \text{ O}_2 \rightarrow 2 \text{ Fe}_2\text{O}_3$$

$$\text{The number of moles of Fe} = (3.0 \text{ kg})\left(1000 \frac{\text{g}}{\text{kg}}\right)\left(\frac{1 \text{ mol}}{55.85 \text{g}}\right)$$

$$= 54 \text{ mol of Fe}$$

$$\text{moles of Fe}_2\text{O}_3 = (54 \text{ mol Fe})\left(\frac{2 \text{ mol Fe}_2\text{O}_3}{4 \text{ mol Fe}}\right) = 27 \text{ mol}$$

To determine the number of grams, we multiply 27 moles of Fe_2O_3 by the molar mass.

$$\text{The molar mass of Fe}_2\text{O}_3 = (2 \times 55.85 \text{ g/mol}) + (3 \times 16.00 \text{ g/mol}) = 159.7 \frac{\text{g}}{\text{mol}}$$

$$27 \text{ mol} \times 159.7 \frac{\text{g}}{\text{mol}} = 4300 \text{ g Fe}_2\text{O}_3 \text{ (2 significant figures)}$$

10. (1) MX_2
    (2) The mole ratio $X_2:MX_2$ is 1:2 so 0.342 moles of X_2 requires a minimum 0.684 moles of MX_3.
    (3) 8 MX_2 theoretically yields 8 MX_3 but only 6 MX_3 were obtained at the end of the reaction. The actual yield is therefore 75%.

11. 2.31 g

2.000 gallons is equal to 7568 mL which has a density of 1.018 g/mL, The mass of the solution therefore is:

$$(1.018 \text{ g/mL})(7568 \text{ mL}) = 7704 \text{ g}$$

Of this mass, 7704 g, 300.ppm are silver ion. To calculate how many grams of silver might be reclaimed with no loss, we set up the following equation.

$$\left(\frac{1}{1 \times 10^6 \text{ ppm}}\right)(300. \text{ ppm})(7704 \text{ g}) = 2.31 \text{ g (3 significant figure)}$$

12. 0.161 g filter paper

The equation tells us that 1 mole of C reduces 4 moles of Ag ion to metallic silver.

Therefore, moles of carbon required is equal to

$$\left(\frac{2.31 \text{ g Ag}}{107.9 \text{ g/mol}}\right)\left(\frac{1 \text{ mol C}}{4 \text{ mol Ag}}\right) = 0.00535 \text{ mol carbon}$$

We need a piece of filter paper that will contain at least 0.00535 moles carbon. The percentage of carbon in the cellulose is

$$\%C = \frac{72.06 \text{ g}}{180.2 \text{ g}} \times 100 = 39.99\%$$

To calculate the mass of filter paper needed we can convert the number of moles of carbon into grams and divide this value by the amount of carbon in the paper.

$$\text{paper mass} = (0.00535 \text{ mol C})(12.01 \text{ g/mol})\left(\frac{1 \text{ g filter paper}}{0.3999 \text{ g C}}\right) = 0.161 \text{ g}$$

CHAPTER TEN

Modern Atomic Theory and the Periodic Table

SELECTED CONCEPTS REVISITED

A brief summary of the contributions by some scientists covered in this chapter is:

Max Planck stated that energy is not released continuously, but in quanta.
Niels Bohr proposed that electrons are found in quantized energy levels or orbits.
Louis de Broglie suggested that all objects have wave properties.
Erwin Schroedinger's mathematical model of an electron's wave properties led to the proposal of electrons being found somewhere within a given area called an orbital, and not along a specific path or orbit.

Electron configurations designate in which orbital each electron of the atom is found. Use the periodic table to help you remember the order of filling orbitals. Divide the periodic table into the s-block, d-block, p-block, and f-block as shown below. Notice the blocks follow the natural divisions of the periodic table as written (this is no accident!).

Here is a summary of some things to remember:

Similar orbitals are half-filled before electrons are paired up in those orbitals. (E.g. the three 2p orbitals in n each "house" one electron. For an O atom, two of the 2p orbitals have a single electron and one has a pair of electrons as opposed to one being empty and the other two having two electrons each.)

The d orbitals are filled after the next higher s orbitals. For example, the 3d orbitals are filled after the 4s orbitals. The 5d orbitals are filled after the 6s orbitals.

The f orbitals are two less than the s orbitals. That is, the 4f orbitals are filled after the 6s orbitals.

orbital diagram $\uparrow\downarrow$ $\uparrow\downarrow$ $\uparrow\downarrow$ $\uparrow\downarrow$ $\uparrow$

electron configuration $1s^2 2s^2 2p^5$

(The next page of this Study Guide shows a Periodic Table with the s, p, d, and f blocks marked)

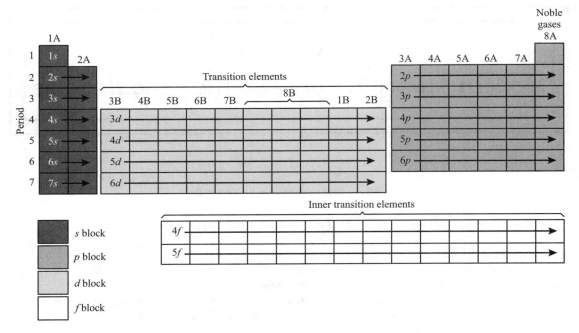

COMMON PITFALLS TO AVOID

You can easily reduce electron-configuration errors by checking to ensure that the sum of the superscripts is equal to the total number of electrons in the atom. For example, F has an electron configuration of $1s^2 2s^2 2p^5$. Sum of the superscripts $= (2 + 2 + 5) = 9$ and an atom of F has 9 electrons so you at least know you assigned the correct number of electrons.

Remember that the 3d level is filled after the 4s level, i.e., $4s^2 3d$ (Not $4s^2 4d$)

$3s^2 3p^6 4s$ NOT $3s^2 3p^6 3d$ – You can avoid making this mistake if you use your periodic table to help you with electron configuration. As you go from left to right across the table until you reach the end of a row and then go to start of the next row, you proceed from the p-block (at the right of the table) to the s-block (at the far left of the table). The d-block comes after the s-block.

SELF-EVALUATION SECTION

1. According to the Bohr model of the hydrogen atom,

    (1) what happens to the electron when it absorbs energy?

    (2) when an electron loses energy, what happens to that excess energy?

2. Fill in the blanks

    (1) The current theory that best explains electron behavior is called the _____ _____ theory.

    (2) This theory states that it is impossible to know the exact position of an electron at any instant of time; instead the region of space where it is most probable to find an electron is called an _____ _____.

3. When n = 1, a total of (1) _____ electrons will fill the (2) _____ sublevel.

    When n = 2, a total of (3) _____ electrons will fill the (4) _____ (the outermost) sublevel.

    Given the symbol $4s^2$, the 4 represents (5) _____, the s represents

    (6) _____, and the 2 represents (7) _____.

4. Review the basic rules and energy level order regarding the state of electrons in atoms.
    a. In the ground state (lowest energy state) of an atom, electrons tend to occupy orbitals of the lowest possible energy.
    b. Each orbital may contain a maximum of two electrons (with opposite spins).
    c. The energy level order is.

        $1s, 2s, 2p, 3s, 3p, 4s, 3d$

    The maximum number of electrons that can be found in any main energy level can be determined by a formula involving the term n where n is the (1) _____ . When $n = 1$, the total number of electrons will be (2) _____ , which will fill the sublevels (3) *s, p, d, f.* For $n = 2$, the total number of electrons will be (4) _____ , which will fill the sublevels (5) *s, p, d, f.* Look at the following symbol and describe its meaning:

        $4s^2$

    The number 4 represents (6) _____ , the small letter s represents (7) _____ , and the superscript 2 represents (8) _____ .

    Now we can write electron configurations. For example, lithium has atomic number 3. Therefore, the correct way of writing the electron configuration would be Li $1s^2 2s^1$.

    Try the configurations for   (9) Nitrogen   Atomic number 7
                                  (10) Argon   Atomic number 18
                                  (11) Sodium   Atomic number 11
                                  (12) Iron   Atomic number 26
                                  (13) Calcium   Atomic number 20

    Write your answers here.

Determine the atomic number for the element represented by the following electron configurations.

(14) $1s^2 2s^2 2p^6 3s^2 3p^2$ _____

(15) $1s^2 2s^2 2p^4$ _____

(16) $1s^2 2s^2 2p^6 3s^2 3p^6 4s^2 3d^7$ _____

(17) $1s^2 2s^2 2p^6 3s^2 3p^6 4s^2 3d^1$ _____

(18) $1s^2 2s^2 2p^6 3s^2 3p^6 4s^2$ _____

5.   Atomic structures may also be represented by orbital diagrams. Using this method, diagram the electron structure for the following. For the first three, circle the orbitals containing the valence electrons.

(1) $_6C$ (4) $_{23}V$

(2) $_{19}K$ (5) $_{33}As$

(3) $_{10}Ne$ (6) $_{17}Cl$

Write your answers here.

6.   Indicate whether the statement made is true or false, or fill in the appropriate blank.

(1) T/F   Known elements are arranged in a table according to increasing atomic mass.

(2)   A horizontal group of elements in the periodic table is called a _____.

(3)   A vertical group of elements in the periodic table is called a _____.

(4) T/F   Elements in the same row have similar chemical properties because they all have the same number of electrons in their outermost energy level.

(5) T/F   All transition elements are metals.

(6)   The distinguishing electrons for the transition elements are filling the ____ and ____ shells.

7.   Using the periodic table in Chapter 10 of the textbook, find the following items of information about each element.

	Symbol	Atomic Number	Atomic Mass	No. of Valence Electrons
(1) Phosphorus	_____	_____	_____	_____
(2) Fluorine	_____	_____	_____	_____
(3) Mercury	_____	_____	_____	_____
(4) Cesium	_____	_____	_____	_____

8. Identify the element represented by the orbital diagram.

(1) ⇅ ⇅ ⇅ ⇅ ↑

(2) ⇅ ⇅ ⇅ ⇅ ⇅ ⇅ ⇅ ⇅ ⇅ ⇅ ↑ ↑ ↑ ▢ ▢

9.

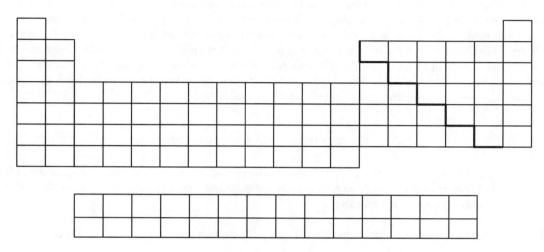

(1) For each of the following, label the appropriate row or column on the periodic chart.
   (a) alkali metals
   (b) halogens
   (c) transition metals
   (d) noble gases
   (e) alkaline earth metals

(2) On the periodic chart, draw an arrow on Group 4A showing the direction of <u>increasing</u> metallic characteristics.

Challenge Problem

10. After reviewing examples in the text for a few minutes, try your hand at writing the complete electronic configuration ($1s$, $2s$, etc.) for (1) $_{21}Sc$ (2) $_{29}Cu$ (3) $_{40}Zr$.

Write your answers here.

In Chapter 10 the model of the atom is modified to our modern ideas. The Bohr model made the first attempt to locate the electrons within the atom. Unfortunately, the concept of orbits could not be used beyond hydrogen and a more accurate model was soon devised. In modern theory of the atom the actual path of the electron is not known. Instead, electrons occupy regions of space known as orbitals. This results in an atomic model of a small dense nucleus, surrounded by a series of electron clouds. The knowledge of energy levels and sublevels allows us to write electron configuration. The special stability of the noble gases can be attributed to the stability associated with 8 electrons in the valence electron level (i.e, the s and p orbitals of the outer energy level are filled). The material in Chapter 10 has also introduced us to a most important practical tool, the periodic table. The information contained in the complete long form of the periodic table is used by scientists everywhere during their day-to-day work. As you continue on in science, you will find yourself referring to the table to calculate molar masses of chemicals, to look up physical properties, and to determine theoretically possible chemical formulas.

ANSWERS TO QUESTIONS AND SOLUTIONS TO PROBLEMS

1.   (1)   In the Bohr model, when an electron absorbs energy, it makes a quantized jump to a higher energy level.
      (2)   When an electron loses energy and makes the quantized jump to a lower energy level, the excess energy is given off as light energy.

2.   (1) quantum mechanics (or wave mechanics)   (2) electron cloud

3.   (1) 2   (2) s   (3) 6   (4) p   (5) principal energy level
     (6) sublevel (s, p, d, f)   (7) the number of electrons in that sublevel

4.   (1) principal energy level
    (2) 2   (3) s (4) 8   (5) s, p (6) principal energy level
    (7) Type of sublevel   (8) Number of electrons in that sublevel
    (9) $1s^22s^22p^3$ (10) $1s^22s^22p^63s^23p^6$ (11) $1s^22s^22p^63s^1$
    (12) $1s^22s^22p^63s^23p^64s^23d^6$ (13) $1s^22s^22p^63s^23p^64s^2$
    (14) 14   (15) 8   (16) 27   (17) 21   (18) 20

5.   (1) C

1s	2s	2p		
↑↓	↑↓	↑	↑	☐

(2s and 2p circled as valence)

    (2) K

1s	2s	2p			3s	3p			4s
↑↓	↑↓	↑↓	↑↓	↑↓	↑↓	↑↓	↑↓	↑↓	↑

(4s circled)

    (3) Ne

1s	2s	2p		
↑↓	↑↓	↑↓	↑↓	↑↓

(2s and 2p circled as valence)

    (4) V

1s	2s	2p			3s	3p			4s
↑↓	↑↓	↑↓	↑↓	↑↓	↑↓	↑↓	↑↓	↑↓	↑↓

3d				
↑	↑	☐	☐	☐

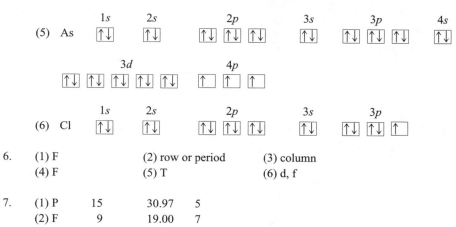

(5) As

	1s	2s	2p			3s	3p			4s

(6) Cl

	1s	2s	2p			3s	3p		

6. (1) F   (2) row or period   (3) column
     (4) F   (5) T   (6) d, f

7. (1) P   15   30.97   5
   (2) F   9   19.00   7
   (3) Hg   80   200.6   2
   (4) Cs   55   132.9   1

8. (1)   Fluorine   (2)   Titanium

9.

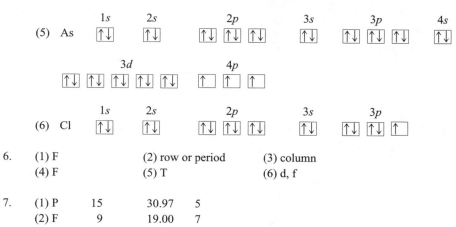

10. (1) $_{21}$Sc $1s^2$ $2s^2$ $2p^6$ $3s^2$ $3p^6$ $4s^2$ $3d^1$
   (2) $_{29}$Cu $1s^2$ $2s^2$ $2p^6$ $3s^2$ $3p^6$ $4s^1$ $3d^{10}$
   (3) $_{40}$Zr $1s^2$ $2s^2$ $2p^6$ $3s^2$ $3p^6$ $4s^2$ $3d^{10}$ $4p^6$ $5s^2$ $4d^2$

CHAPTER ELEVEN

Chemical Bonds: The Formation of Compounds from Atoms

SELECTED CONCEPTS REVISITED

Ionization energy is the energy required to remove an electron from an atom in the gaseous state. Therefore the larger the ionization energy, the more difficult it is to remove an electron. Remember that if you are removing an electron, what is left behind is positively charged. Also, the remaining electrons now feel a stronger pull to the nucleus (since there is now more positive charge than negative charge) and so it is harder to pull away subsequent electrons.

Lewis structures show valence electrons. We try to show the approximate distribution of the electrons in the valence orbitals, that is, whether they are paired or not.

Ionic bonds are the result of electrostatic interaction (that is, opposite charges attract). Ions (charged species) are formed when electrons are transferred.

Covalent bonds are the result of a pair of electrons being shared between two atoms. When one atom involved in a covalent bond has a stronger attraction for the shared pair of electrons than the other atom, the result is a polar covalent bond. That is, the electrons are being shared but not equally and because they are more likely to be found nearer the more electronegative atom, that atom will have a partial negative charge. The less electronegative atom will have a partial positive charge.

Notice that the more electronegative elements are the nonmetals. One rationale for the trend in electronegativity is that the smaller the atom, the closer to the nucleus the shared electrons will be and so the more the nucleus is able to attract the electrons. Fluorine is the most electronegative element.

A molecule cannot be polar unless it contains polar bonds. However, a molecule that contains polar bonds is not necessarily polar. To determine if a molecule is polar, you need to look at the shape of the molecule. If the polar bonds are able to cancel each other out, the molecule is not polar. (You can imagine these polar bonds as miniature tug-of-wars. If there can be a winner, the molecule is polar.)

VSEPR theory allows you to predict the shape of a molecule. The idea is that lone pairs and bonds all contain electrons and like charges repel so these areas of negative charges repel each other such that the bonds and lone pairs try to get as far apart from each other as possible. (If you use a balloon to represent a pair of electrons, whether a bonding pair or a nonbonding pair, and tie four balloons together, you will see the balloons point to the four corners of a tetrahedron.)

The number (and therefore arrangement) of electron pairs determine the **geometric shape** of the molecule; the position of the atoms around the center atom determines the **molecular shape** of the molecule. The latter is generally what most people think of when asked for the shape of the molecule. Be sure to clarify what you are being asked for!

COMMON PITFALLS TO AVOID

Electrons are negatively charged particles. If atoms gain electrons, they become negatively charged, if they lose electrons, they become positively charged.

Do not confuse valence electrons with electrons on an atom. Electrons in the inner shells are generally not involved in chemical reactions. Only the outermost electrons (the valence electrons) are of concern in Lewis structures.

When drawing Lewis structures of covalent species, it is helpful to assign the lone pairs to the outer atoms first. Then if the electrons "run out" before the central atom has its octet, simply convert a lone pair on an outer atom into a bonding pair between that atom and the central atom. Double bonds and extra electrons are often around the central atom. Also, remember it is often easy to find a mistake in the Lewis structure of a covalent species. If each atom does not have eight electrons (other than H) and the total number of electrons is not equal to the total number of valence electrons of those atoms, then you have made a mistake.

Do not forget to use VSEPR to determine the shape of the molecule **before** trying to decide if the molecule is polar.

SELF-EVALUATION SECTION

1. Fill in the blank space or circle the appropriate response.
    When a neutral atom loses an orbital electron, it becomes a (1) positively/negatively charged ion. The energy

    required to remove a mole of electrons from a mole of atoms is called the (2) _____ energy and is a
    relatively (3) low/high value for Group I and Group II metals and (4) low/high for non-metallic elements. The ionization energy for the noble gas elements is especially (5) high/low, indicating that eight electrons in the valence shell of an atom is a very stable structure. If you list the ionization energies of the elements from a group in the periodic table, the element from the top of the group has a (6) higher/lower ionization energy than the element at the bottom of the group. Two factors account for this experimental observation. As you go down a group, the electron being removed is (7) farther from/closer to the nucleus, and the increasing number of filled electron orbitals shields the valence electrons from the positive nucleus. The valence electrons are shown in the Lewis structures. We will use Lewis structures in the next section to illustrate the concept of forming chemical bonds.

    As mentioned above, when a neutral atom (8) loses/gains an electron it becomes positively charged. A

    positively charged ion is also called a (9) _____ . For example, a potassium atom has 19 protons

    and 19 electrons. A potassium ion has a charge of $+1$, which means the ion has (10) _____

    electrons instead of 19, as in the neutral atom. Conversely, a negatively charged ion, which is called an

    (11) _____ , is formed when a neutral atom (12) loses/gains electrons. A chlorine atom has
    17 protons and 17 electrons and becomes an anion by (13) gaining/losing an electron. A chloride ion has a stable

    outer shell of (14) _____ electrons. A sodium atom in close proximity to a chlorine atom can
    also reach a stable outer shell of eight electrons by losing one electron. Thus, both sodium and chlorine reach a stable electron structure by the process of electron transfer. The metallic elements attain a stable structure by (16) gaining/losing electrons; the nonmetallic elements attain a stable structure by (17) gaining/losing electrons.

2. Using Lewis structures, illustrate how the following compounds are formed. Some are electron-transfer problems and some are electron-sharing problems. Use the information listed below or a periodic table.

Element	Group
Zinc	2B
Potassium	1A
Hydrogen	1A
Iodine	7A
Oxygen	6A
Phosphorus	5A
Chlorine	7A
Sulfur	6A
Carbon	4A

(1) zinc iodide (ZnI_2)

(2) potassium oxide (K_2O)

(3) phosphorus trichloride (PCl_3)

(4) hydrogen sulfide (H_2S)

(5) carbon dioxide (CO_2)

(6) sulfur trioxide (SO_3)

3. Write out the Lewis structure for HNO_2, nitrous acid. You may use a periodic table and follow problems in the text as a guide.

4. Draw Lewis structures for the following:

    (1) MnO_4^- (2) CO_3^{2-} (3) NH_4Cl

    (4) PO_4^{3-} (5) ClO^- (6) H_2O

5. If we examine many different compounds for the kinds of chemical bonds that hold them together, we will generally find two types. The two types of chemical bonds are called (1) _____ and the (2) _____ bond. When a transfer of electrons takes place from one atom to another, an (3) _____ bond is formed.

    A cation and an anion will form an (4) _____ bond since oppositely charged particles (5) attract/repel each other. Metallic elements tend to form ionic bonds when combining with the nonmetals, as we saw in the previous section.

    The predominant type of chemical bond is the (6) _____ bond. This type of bond occurs in the hydrogen molecule and develops as a result of each hydrogen atom contributing (7) one/two electron(s) to form the bond. The $1s$ electron orbitals of the hydrogen atoms overlap and pair to form a stable hydrogen molecule. There is a strong tendency for the hydrogen molecule to form from two individual atoms since in the molecule each (8) _____ charged electron is attracted to two (9) _____ charged nuclei.

    The covalent bond is usually indicated by a dash mark (—). A single dash means (10) _____ pair of electrons and a double dash means (11) _____ pairs of electrons.

6. Given the following table of electronegativity values, indicate which of the listed binary compounds has polar covalent bonds. Also, calculate the difference in electronegativity values for one of the covalent bonds in each molecule.

    Electronegativity Values

H	2.1	Br	2.8	Se	2.4
B	2.0	Cl	3.0	Te	2.1
P	2.1	F	4.0	N	3.0
O	3.5	S	2.5	Na	0.9

    For example, the compound HF has one covalent bond. Is the bond between H and F polar? Yes, there is a significant difference in electronegativity values of $4.0 - 2.1 = 1.9$. Examine the remaining compounds in the same manner.

	Polar Covalent Bond (yes or no)	Electronegativity Value Difference
(1) NCl_3	_____	_____
(2) BrCl	_____	_____
(3) Na_3P	_____	_____
(4) H_2Se	_____	_____
(5) H_2O	_____	_____
(6) Br_2	_____	_____
(7) OF_2	_____	_____
(8) PH_3	_____	_____
(9) H_2Te	_____	_____
(10) NH_3	_____	_____
(11) H_2S	_____	_____

7.   In the following, which of each pair will be larger?

(1) Cl^- and Cl   (3) Na^+ and Na
(2)   Al and Al^{3+} (4) Fe^{2+} and Fe^{3+}

8.   (1)   Draw the Lewis structure for SiO_2.
     (2)   Are the bonds in SiO_2 polar covalent?
     (3)   Is the molecule SiO_2 polar or non-polar?

9.   Predict the type of bond that would be formed between the following pairs of atoms. Use Table 11.5 in the text.

(1)   Li & I   (2)   C & H   (3)   Al & N
(4)   Cs & P   (5)   Se & S   (6)   Ba & C

RECAP SECTION

Chapter 11 presents you with a great deal of useful information about the formation of chemical compounds from individual atoms. You now know how to describe chemical bonding in terms of electron transfer or electron sharing. And by using the concept of electronegativity, you can be more precise about the kind of electron sharing that takes place in covalent bonding. The concept of electronegativity is a most valuable tool in chemistry. The general theme of the unit has been to examine the formation of compounds and to describe the forces of attraction in compounds. The last sections in the chapter discuss the three-dimensional shapes of molecules and how to use a model called "valence shell electron pair repulsion" to predict molecular shape from Lewis structures.

ANSWERS TO QUESTIONS AND SOLUTIONS TO PROBLEMS

1. (1) positively   (2) ionization   (3) low   (4) high
    (5) high   (6) higher   (7) farther from   (8) loses
    (9) cation   (10) 18   (11) anion   (12) gains
    (13) gaining   (14) eight   (15) losing   (16) gaining

2. (1) Zinc is a Group 2B element, which means that it has two valence electrons in its outer shell. Iodine is a Group 7A element, which means that it has seven valence electrons in its outer shell. Zinc loses one electron to each of the iodine atoms and becomes a +2 charged cation. Each iodide ion thus has a −1 charge.

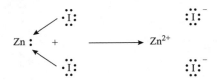

(2)

(3)

(4)

(5) In order to write a Lewis structure for the molecule CO_2, which will have eight electrons around each atom, we must use double bonds between C and each oxygen atom. Each double bond consists of four electrons – two from the C atom and two from each oxygen atom.

The problem involves the formation of double bonds. Since one pair of electrons is one bond, there are two bonds between the two oxygen atoms and the carbon atom. Only in this manner will each atom have eight electrons in its outer shell.

(6)

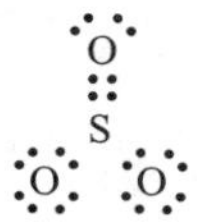

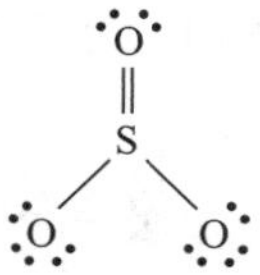

The Lewis structure for SO_3 involves one double bond and two single bonds in order to place eight electrons around each atom.

3. The first step was to count the total number of valence electrons in the molecule.
    # valence electrons $= 1 + 5 + 6 + 6 = 18$

    Next write the skeletal structure of the atoms and connect them with a single bond (two electrons). How do we know the arrangement of the atoms? One clue is that HNO_2 is an acid, which suggests the H is bonded to an O.

    H:O:N:O or H-O-N-O

    Then figure out how many electrons are left to be assigned.

    Started with 18 electrons and used two for each bond shown. $18-6 = 12$. 12 electrons left

    Now distribute pairs of electrons around each atom (nonbonding pairs) resulting in noble gas configurations. Note that H only needs a total of 2 electrons to achieve "noble gas status". Most of the rest of the elements need 8 electrons. (Hint, assign nonbonding pairs to "outer" elements first.)

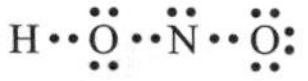

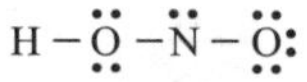

    The twelve remaining electrons are used (i.e., all 18 have now been assigned), but not all the atoms are satisfied. The H has two electrons (good), the oxygens each have 8 electrons (good), but the nitrogen only has 6 electrons around it; it wants two more.

    So the penultimate step is to shift a nonbonding pair of electrons from an outer atom to make a double bond. In other words, share a pair of electrons with the central atom.

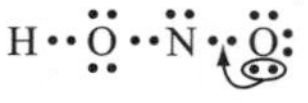

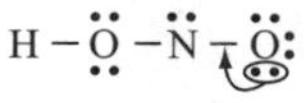

    And the last step is to always double-check! 18 electrons total with 8 electrons around each atom (except for H).

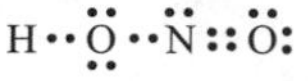

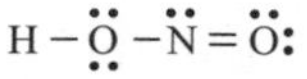

4. (1) 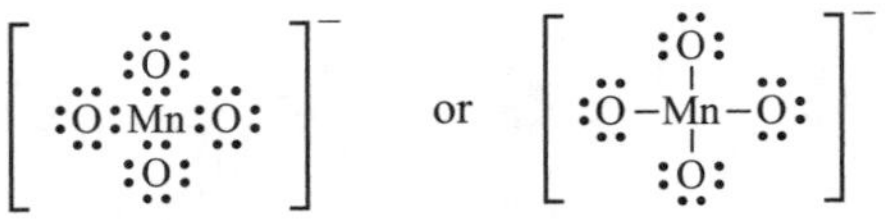 [:O:
:O:Mn:O:
:O:]⁻ or [:O:
:O̶—Mn—O̶:
:O:]⁻

(2) 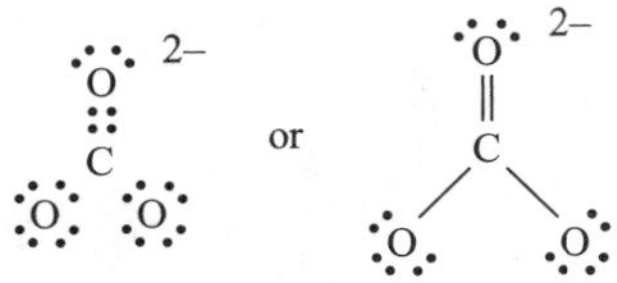 :O: 2−
C
:O: :O: or   :O: 2−
‖
C
O: :O:

(3) 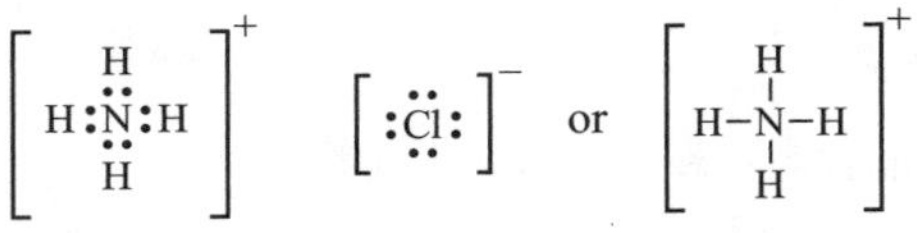 [H
H:N:H
H]⁺ [:Cl:]⁻ or [H
H—N—H
H]⁺

(4) 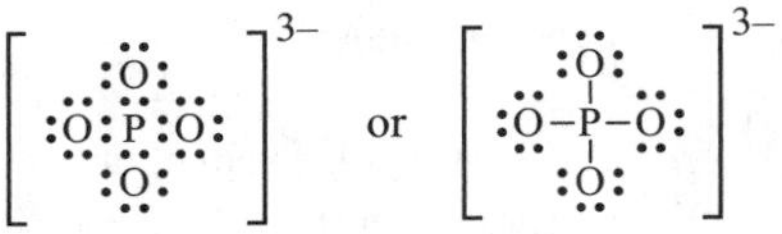 [:O:
:O:P:O:
:O:]³⁻ or [:O:
:O—P—O:
:O:]³⁻

(5) 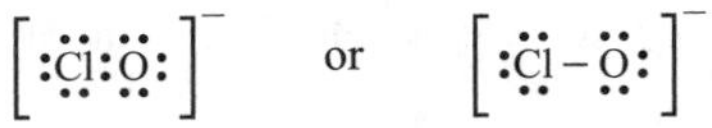 [:Cl:O:]⁻   or   [:Cl—O:]⁻

There are 4e⁻ from the carbon, 6e⁻ each from the oxygen and 2 additional (negative charge on the ion is −2). So we have to distribute 24e⁻ between the central carbon and the 3 oxygen atoms. This means that one of the bonds needs to be shown as a double bond.

(6) 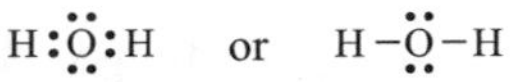 H:O:H   or   H—O—H

5. (1) ionic   (2) covalent   (3) ionic
    (4) ionic   (5) attract   (6) covalent
    (7) one   (8) negatively   (9) positively
    (10) one   (11) two

6.
Compound	Polar Covalent Bond (yes or no)	Electronegativity Value Difference
(1) NCl_3	no	0
(2) BrCl	yes	0.2
(3) Na_3P	yes	1.2
(4) H_2Se	yes	0.3
(5) H_2O	yes	1.4
(6) Br_2	no	0
(7) OF_2	yes	0.5
(8) PH_3	no	0
(9) H_2Te	no	0
(10) NH_3	yes	0.9
(11) H_2S	yes	0.4

7. (1) Cl$^-$ will be larger   (3) Na will be larger
    (2) Al will be larger   (4) Fe^{2+} will be larger

8. (1) SiO$_2$ has $(4 + 6 + 6) = 16$ valence electrons. Its Lewis structure is

$$\ddot{\underset{..}{O}} = \ddot{Si} = \ddot{\underset{..}{O}}$$

    (2) Both sets of double bonds between Si and O are polar because of the difference in the electronegativities of Si and O.

    (3) SiO$_2$ is a non-polar molecule. Although each of the double bonds is polar, the molecule is linear, so the polarities negate each other.

9. (1) polar covalent   (2) polar covalent   (3) polar covalent
    (4) polar covalent   (5) nonpolar covalent   (6) ionic

WORD SEARCH 1

In the given matrix of letters, find the terms that match the following definitions. The terms may be horizontal, vertical, or on the diagonal. They may also be written forward or backward. Answers are found at the end of Chapter 20.

1. A subatomic particle with a charge of $+1$.
2. The mass of an object divided by its volume.
3. The basic building block of matter that cannot be broken down into simpler substances by ordinary chemical changes.
4. An electrically charged atom or group of atoms.
5. The central part of an atom
6. An element that is ductile and malleable.
7. The abbreviation for the name of an element.
8. The chemical law concerning the occurrence of the chemical properties of elements.
9. State of matter that is least compact.
10. Atoms of an element having the same atomic number but different atomic masses.
11. A subatomic particle with a charge of -1.
12. A negatively charged particle.
13. A small, uncharged individual unit of a compound.
14. Matter having uniform properties throughout.
15. The smallest particle of an element that can enter into a chemical reaction.
16. The relative attraction that an atom has for the electrons in a covalent bond.
17. A molecule with a separation of charge.
18. A subatomic particle that is electrically neutral.
19. The energy required to remove an electron from an atom.
20. A substance composed of two or more elements combined in a definite proportion by mass.
21. A solid without definite crystalline form.
22. A cloud-like region around the nucleus where electrons are located.
23. The metallic elements characterized by increasing numbers of d and f electrons in an inner shell.
24. Metric unit of length.

H	C	O	M	P	O	U	N	D	D	I	S	O	T	O	P	E	S
O	X	C	S	Z	K	L	O	Y	P	M	N	Q	R	B	L	C	T
M	P	R	U	D	K	Q	T	V	F	V	G	M	W	E	Z	T	Y
O	I	Z	E	O	R	V	O	M	T	S	E	O	C	U	A	L	G
G	U	A	L	T	E	I	R	O	R	E	S	T	T	L	A	C	I
E	I	F	C	S	U	P	P	W	L	L	R	A	B	A	I	T	U
N	G	O	U	Q	E	G	Y	O	H	O	J	Y	G	R	N	O	P
E	E	E	N	E	U	T	R	O	N	P	A	M	V	E	E	S	E
O	L	B	U	I	S	W	H	E	J	I	M	N	M	T	K	O	R
U	S	O	V	T	Z	U	G	J	F	D	V	E	C	L	D	E	I
S	I	E	I	R	K	A	O	A	B	K	L	T	U	M	L	E	O
K	N	G	S	D	T	M	T	H	V	E	P	R	S	D	D	A	D
L	L	A	T	I	B	R	O	I	P	A	D	O	I	M	E	R	I
O	E	W	V	S	L	H	F	L	O	R	G	E	U	Q	N	V	C
B	O	I	O	N	N	A	P	A	E	N	O	L	T	W	S	Z	L
M	T	V	M	E	O	S	I	D	N	C	E	M	N	O	I	N	A
Y	L	S	A	T	R	T	Z	M	A	B	U	N	A	L	T	U	W
S	P	J	D	A	T	C	I	O	T	M	J	L	E	K	Y	V	Z
A	E	O	R	B	C	M	R	P	E	B	U	N	A	L	T	U	W
B	H	T	L	C	E	A	H	T	M	X	Y	U	T	T	G	G	C
G	D	I	U	F	L	Q	A	O	F	E	S	N	X	G	A	Y	V
S	T	N	E	M	E	L	E	N	O	I	T	I	S	N	A	R	T
A	M	G	W	U	P	Q	N	S	W	D	L	E	R	V	K	C	B
H	I	V	R	E	T	E	M	O	R	D	Y	H	R	W	E	J	F

CHAPTER TWELVE

The Gaseous State of Matter

SELECTED CONCEPTS REVISITED

An important assumption of the kinetic-molecular theory is that gas molecules move in straight lines and collide with the walls of the container. Pressure, whether it's water pressure or gas pressure, is the force exerted on a unit area by a substance. As gas molecules collide with the container walls, they exert a certain pressure. What will happen to the pressure of a gas sample if more gas molecules are placed in a container? More molecules mean more collisions with the walls. Therefore, the pressure will rise. If the temperature and volume of the container are kept constant, there is a direct relationship between pressure and number of gas molecules. Thus, doubling the amount of gas will double the pressure. Reducing the amount of gas to one-third the original quantity will reduce the pressure correspondingly. Refer to Figure 12.4 in the text.

The kinetic molecular theory basically says that ideal gases are composed of tiny massless particles having no attraction for each other that move in straight lines and lose no energy when they collide with the walls of the container.

The average kinetic energy of a gas is dependent on its temperature and not on the size of the particle. The mass of the particle affects its velocity not its energy.

The process of effusion is movement of a gas through a small opening. A comparison of the rate at which two gases at the same temperature will effuse shows that the lighter gas effuses faster. Think about this and it makes sense. If two gases are at the same temperature then they have the same average kinetic energy so the lighter gas moves faster and therefore should pass through an opening faster.

The pressure exerted on a container is due to the collisions of the gas molecules on the container. More collisions lead to higher pressures. The number of collisions is affected by the number of particles present, their speed (once they hit one wall, how long before they reach another wall), and the volume of the container (how far do the particles have to travel before they reach another wall). So the pressure of a gas is dependent on the moles of gas present, the temperature of the gas, and the volume of the container. This is basically the information contained in the gas laws.

What happens to a gas sample if the pressure is not constant and the volume cannot expand? Visualize a closed empty can placed on a fire. The temperature of the gas increases; kinetic energy and pressure increase. The gas cannot expand so the pressure increases to the point at which the mechanical strength of the can cannot withstand the high pressure, and the can explodes.

The kinetic molecular theory assumes that the particles of a gas are independent of each other. In a mixture of gases, we can treat each gas in the mixture as though it were the only gas present. So to find the total pressure exerted on the container, we can simply find the pressure exerted by each gas and add them. That is, the total pressure is the sum of the partial pressures. Alternatively, if you were given the total number of moles of all the gases present, we simply find the total pressure and not really worry about the fact that there are different gases present.

Unlike solids and liquids, the density of a gas changes significantly with changes in temperature or pressure. For a given mass of a gas, its volume is highly dependent on its temperature and pressure and therefore its density is affected. So although we can generalize and say that the density of liquid water is 1 g/ml, if we give the density of steam, we would need to specify at what temperature and pressure the density was measured. The density of gases generally uses units of g/L.

When gases react at constant temperature and pressure, the coefficients of a balanced equation can refer to either the volumes of the gas used or the moles of gas used. This again falls back on the fact that regardless of the identity of a gas, at a specific temperature and pressure, one mole of a gas will occupy a specific volume. At STP, that volume (the molar volume) is 22.4 dm^3 (liters).

Ozone is decidedly contrary. It is good and bad. It all depends on where it is found. We desire to reduce air pollution (e.g. from vehicles and factories), and therefore ozone, in the lower atmosphere because it can potentially be fatal in high concentrations. On the other hand, we want ozone in the upper atmosphere (specifically the stratosphere) because it absorbs uv radiation. (If only we could figure out a way to transport it from the lower atmosphere to the stratosphere.)

COMMON PITFALLS TO AVOID

Do not get confused between effusion and diffusion. Effusion is the movement of a gas through a small opening. The gas flows from an area of high pressure to an area of low pressure. Diffusion is the spontaneous mixing of gases.

Be careful when using the ideal gas constant R in calculations. Make sure that the units of R that you use are consistent with your other values. Also be extremely careful to check the units of temperature used in the calculations! Temperature must be in Kelvin for all calculations involving gases.

Room temperature and STP are not the same. STP is an abbreviation for standard temperature and pressure. Standard temperature is 273 K (or 0°C). Room temperature is often assumed to be 298 K (or 25°C). Standard pressure is 1 atm.

PV = nRT is not the answer to all of your problems! Whenever you do a calculation also try to see if the answer makes sense qualitatively. However, remembering the ideal gas equation can help you remember certain relationships. From PV = nRT you know that $P \propto 1/V$ (assuming the rest of the variables are constant), so if pressure is increased, the volume must have decreased. You can often use logic to get an idea of the relative quantity of the answer even before you start to plug numbers into your calculator.

For most calculations, using 273k instead of 273.15 to convert between K and °C is acceptable.

SELF-EVALUATION SECTION

1.   The science of chemistry on a quantitative basis began with the systematic study of gas behavior, which may seem somewhat ironic since substances in the gas state are difficult to handle and use for experimental purposes. The fundamental properties of gases were established as compressibility, diffusion, pressure, and expansion. In addition, equal volumes of gases at the same temperature were found to exert identical pressures.

The assumptions of the kinetic-molecular theory for an ideal gas are listed below.

    a.   Gases consist of tiny particles.
    b.   The volume of gas is mostly empty space.
    c.   Gas molecules have no attraction for each other.
    d.   Gas molecules move in straight lines in all directions, undergoing frequent collisions.
    e.   No energy is lost through the collisions of gas molecules.
    f.   The average kinetic energy for molecules is the same for all gases at the same temperature.

Properties – list the letter for one or more assumptions from the above list that describe each of the following properties of gases.

    (1)   Compressibility   _____
    (2)   Diffusion   _____
    (3)   Pressure   _____
    (4)   Expansion   _____
    (5)   Pressure of equal volumes of gases   _____

2.   Boyle's law and Charles' law.

    Given below are several ways of mathematically stating Boyle's law and Charles' law. Place a "B" (for Boyle's) or "C" (for Charles') by each formula.

    (1) $PV = k$ (4) $P_1 V_1 = P_2 V_2$

    (2) $\dfrac{V_1}{T_1} = \dfrac{V_2}{T_2}$ (5) $V \propto T$

    (3) $V \propto \dfrac{1}{P}$ (6) $\dfrac{V}{T} = k$

    The symbol "$\propto$" means "to vary" or "to be proportional to".

3.   Which of the gases in each pair will effuse at the fastest rate according to Graham's law?

    (1) H_2, He   (4) NH_3, CO
    (2) SO_2, Br_2 (5) CO_2, Ne
    (3) N_2, Cl_2 (6) Cl_2, F_2

4.   **Gas law problems**

    (1)   The pressure of a gas in a piston is 2500 torr. The pressure is reduced to 610 torr. The original volume was 1.00 liter. What is the new volume?

    Do your calculations here.

(2)   A certain gas exerted a pressure of 580. torr in a 4.00 liter container. The gas was compressed into a 500. mL gas bottle. What was the new pressure?

Do your calculations here.

(3)   120. mL of a gas are obtained from a reaction at 260.°C. What volume will the gas occupy at 100.°C?

Do your calculations here.

(4)   A 125 mL sample of gas originally at −40.°C was brought to room temperature (25°C). Find the new volume.

Do your calculations here.

(5)   You should also be able to work a combined gas law problem that involves P, V, and T. Try this one.

What is the volume of a gas at STP if the original conditions were 50. mL, 0.75 atm pressure, and 21°C?

Do your calculations here.

5.   Make the best match between the Law/quantity and their definitions/quantities.

(1)   ___ Dalton's Law

(2)   ___ Boyle's Law

(3)   ___ Ideal Gas Law

(4)   ___ Avogadro's Law

(5)   ___ Charles' Law

(6)   ___ Gay-Lussac Law

(7)   ___ Graham's Law

(a)   At constant volume and mass, the P of a gas is proportional to T in Kelvin.
(b)   Equal volumes of two gases at the same P and T will contain equal numbers of particles.
(c)   As the volume of a gas decreases at constant T, its pressure increases.
(d)   The volume of a gas is directly proportional to T in Kelvin at constant P.
(e)   Rate of effusion of a gas is inversely proportional to the square root of its mass.
(f)   The result of the compilation of Boyle's, Charles', Avogadro's and Gay-Lussac's Laws.
(g)   The pressure of a mixture of gases is simply the sum of their partial pressures under the same conditions.

Choose the correct measurement for the following commonly used/referred to quantities.

(8) ___ standard temperature   (9) ___ molar volume   (10) ___ standard pressure

(h) 24.0 L   (i) 76 mmHg   (j) 22.4 L
(k) 273 K   (l) 1 atm   (m) 298 K

6.   The density of a gas is expressed in grams per liter.

$$d = \frac{mass}{volume} = \frac{g}{L}$$

At STP, density may also be calculated from the following equation.

$$density\ at\ STP = \frac{molar\ mass}{22.4\ L/mol} = \frac{g/mol}{liters/mol} = \frac{g}{L}$$

(1)   At STP, what is the density of NO (nitrogen oxide) gas? Use your table of atomic masses.

       Do your calculations here.

(2) The density of HCN (hydrogen cyanide) gas at STP is 1.21 g/L. What volume will 100. g of HCN occupy at STP?

Do your calculations here.

(3) A quick way to prepare acetylene gas in the laboratory is to drop chunks of calcium carbide into water. Old-fashioned miners' lamps used this reaction, as did some residential lighting systems years ago. We will work a Dalton's partial pressure problem from this reaction.

A sample of C_2H_2 (acetylene) gas collected over water at 21°C and 750 torr pressure occupies a volume of 175 mL. Calculate the volume of dry acetylene at STP. The vapor pressure of water at various temperatures is listed in Appendix II of the textbook. Dalton's law states that, in a mixture of gases, each gas exerts its own individual pressure.

$$P_{total} = P_A + P_B + P_C$$

For this particular problem, the total pressure of the moist acetylene collected is made up of two parts – the pressure of C_2H_2 and the pressure of the water vapor. We have to subtract the partial pressure of the water vapor from the C_2H_2 before we correct the C_2H_2 volume to STP.

Do your calculations here.

7. What volume of NO gas at STP can be produced from 7.0 moles of nitrogen dioxide (NO_2) according to the following equation?

$$3 NO_2 + H_2O \rightarrow 2 HNO_3 + NO$$

Do your calculations here.

8. Using the ideal gas equation, calculate the volume that 15 g of H_2 gas at 25°C and 1.2 atm pressure will occupy.

$$PV = nRT$$

Do your calculations here.

9. What volume of NO gas will be produced from 4.00 moles of N_2 and 3.00 moles of O_2 at STP according to the following equation?

$$N_2(g) + O_2(g) \rightarrow 2\,NO(g)$$

Do your calculations here.

Challenge Problems

10. In a movie thriller, the hero is locked in a sealed room. The bad guys allow a mixture of two poisonous gases, arsine (AsH_3 – garlic odor) and cyanogen (C_2N_2 – almond odor), to effuse through a porous opening into the room. If the hero doesn't find an escape route, what will be the fragrance that reaches the nostrils first?

Do your calculations here.

11. The size cylinder known as a 1A cylinder has a volume 43.8 L. How many moles of oxygen are contained in a 1A cylinder at 72°F and 1500 pounds per in^2 pressure (psi)? To convert psi to atmospheres multiply psi by 0.06805 atm/psi. What is the mass of this number of moles of oxygen gas?

Do your calculations here.

12. Use the ideal gas equation to calculate the molar mass of arsine (the garlic odor from problem 10), given the following information.

3.48 g of AsH_3 at 740 torr and 21°C occupies 1.11 L of volume. What is the molar mass of AsH_3?

Do your calculations here.

RECAP SECTION

Chapter 12 is long and contains a great deal of new material. However, you should have sharpened up your problem-solving skills and learned a great deal about the gaseous state. Particularly important is the ability to visualize in equation form a word statement such as Boyle's law and then use the relationship to solve problems. The relationships of observed gas properties to the kinetic-molecular theory of ideal gas behavior, Avogadro's Law, Gay-Lussac's law of combining volumes, and Dalton's Law of Partial Pressures were studied. However, it is important to realize that real gases don't exactly behave as the ideal gas equation predicts because real gases do have finite volume and also exhibit intermolecular attractions, particulary at low temperatures and high pressures. You should feel confident in tackling almost any stoichiometric problem, given an equation and an atomic mass table, if not, review this chapter again. You will use these skills again in later chapters including in solution chemistry.

ANSWERS TO QUESTIONS AND SOLUTIONS TO PROBLEMS

1. (1) Compressibility – b
   (2) Diffusion – d
   (3) Pressure – d
   (4) Expansion – b, c, d
   (5) Pressure of equal volumes of gases – f, e

2. (1) B   (2) C   (3) B   (4) B   (5) C   (6) C

3. (1) H_2 (2) SO_2 (3) N_2 (4) NH_3 (5) Ne   (6) F_2

4. (1) 4.1 liters
   The problem involves pressures and volumes; this means a Boyle's law problem.

   The original volume was 1.00 L, and the original pressure was 2500 torr. The pressure dropped to 610 torr. What happens to the volume of a gas when the pressure goes down? According to Boyle's law, the volume increases. This means we must multiply the original volume by a ratio of pressures, which will give us a larger volume than 1.00 L.

   original volume $\times$ ratio of pressure = new volume

   $$(1.00 \text{ L})\left(\frac{2500 \text{ torr}}{610 \text{ torr}}\right) = 4.1 \text{ L}$$

(2) 4.64×10^3 torr

This time let's approach the problem using Method (B) (algebraic) from the chapter. Organizing the data and converting the units to be the same gives:

$P_1 = 580.$ torr $P_2 = ?$
$V_1 = 4.00$ L $V_2 = 500.$ mL $= 0.500$ L

Using Boyle's Law:

$$P_1V_1 = P_2V_2$$

and dividing both sides of the equation by V_2:

$$\frac{P_1V_1}{V_2} = P_2$$

Now substituting the values from the data table:

$$\frac{(580. \text{ torr})(4.00 \text{ L})}{0.500 \text{ L}} = 4.64 \times 10^3 \text{ torr}$$

(3)   84.0 mL

The problem involves volumes and temperatures, which makes it a Charles' law problem. We know that V and T are directly related as long as T is expressed as a Kelvin or absolute temperature. We first have to convert °C to K. Any time you are asked to solve a gas law problem with a temperature involved, be sure to convert to K.

$260.$°C $+ 273 = 533$ K
$100.$°C $+ 273 = 373$ K

The temperature dropped from 533 K to 373 K. This means the volume must decrease also.

original volume $\times$ ratio of temperatures $=$ new volume

$$(120 \text{ mL})\left(\frac{373 \text{ K}}{533 \text{ K}}\right) = 84.0 \text{ mL}$$

(4)   149 mL

Once again we will use Method (B) (algebraic) from the chapter. Organizing the data and converting the temperatures to Kelvin gives:

$V_1 = 125$ mL $V_2 = ?$
$T_1 = -40.$°C $= 233$ K $T_2 = 25$°C $= 298$ K

This is a Charles' Law problem so,

$$\frac{V_1}{T_1} = \frac{V_2}{T_2}$$

Multiplying both sides of the equation by T_2 gives $\dfrac{V_1T_2}{T_1} = V_2$. Substituting

$$\frac{(125 \text{ mL})(298 \text{ K})}{233 \text{ K}} = 149 \text{ mL}$$

(5)   35 mL

With a problem like this, it is useful to tabulate the data in an organized fashion.

$P_1 = 0.75$ atm $P_2 = 1.0$ atm (standard)
$V_1 = 50.$ mL $V_2 = ?$
$T_1 = 21°C = 294$ K $T_2 = 0°C = 273$ K (standard)

Since P, V, and T are all changing, the combined gas law is used

$$\frac{P_1 V_1}{T_1} = \frac{P_2 V_2}{T_2}$$

Solving of V_2 and substituting data:

$$\frac{P_1 V_1 T_2}{T_1} = V_2 = \frac{(0.75 \text{ atm})(50. \text{ mL})(273 \text{ K})}{(1.0 \text{ atm})(294 \text{ K})} = 35 \text{ mL}$$

5.   (1) g   (2) c   (3) f   (4) b   (5) d   (6) a   (7) e   (8) k   (9) j   (10) 1

Note that 24.0 L is the molar gas volume at <u>room</u> temperature. Room temperature is generally considered to be 298 K (25°C). Standard pressure of 1 atm is the same as 760 mmHg not 76 mmHg.

6.   (1) $1.34 \dfrac{\text{g}}{\text{L}}$

The molar mass of NO is 14.01 g/mol + 16.00 g/mol = $30.01 \dfrac{\text{g}}{\text{mol}}$

$$d = \left(\frac{30.01 \text{ g}}{\text{mol}}\right)\left(\frac{1 \text{ mol}}{22.4 \text{ L}}\right) = 1.34 \frac{\text{g}}{\text{L}}$$

(2)   82.6 L
Solve the equation for volume (V).

$$d = \frac{m}{V}$$

Multiply each side of the equation by "V".

$$(V)(d) = \frac{(m)(V)}{(V)}$$

Now divide each side by d

$$\frac{(V)(d)}{(d)} = \frac{m}{d}$$

We can now substitute the values given for the density and the mass into the equation.

$$V = \frac{(100. \text{ g})(1 \text{ L})}{1.21 \text{ g}} = 82.6 \text{ L HCN}$$

(3)   156 mL

Since the acetylene gas was collected over water, we have to subtract the partial pressure of water vapor at 21°C to find the partial pressure of acetylene alone. Then, we can find the volume at STP.

$P_{total} = P_{C_2H_2} + P_{H_2O}$
750 torr $= P_{C_2H_2}$ + 18.6 torr (Appendix II)
$P_{C_2H_2}$ = 750 torr − 18.6 torr = 731 torr

Establish a table

P_1 = 731.4 torr	P_2 = 760. torr (standard)
V_1 = 175 mL	V_2 = x mL
T_1 = 21°C = 294 K	T_2 = 0°C = 273 K (standard)

Using the ratio technique, we see that pressure is increased and the temperature is reduced, both of which reduce the volume.

$$(175 \text{ mL})\left(\frac{731 \text{ torr}}{760. \text{ torr}}\right)\left(\frac{273 \text{ K}}{294 \text{ K}}\right) = 156 \text{ mL}$$

7.   52 L

The number of moles of starting substance is 7.0 moles NO_2.

Calculate the moles of NO, using the mole-ratio method.

$$(7.0 \text{ mol } NO_2)\left(\frac{1 \text{ mol NO}}{3 \text{ mol } NO_2}\right) = 2.3 \text{ mol NO}$$

Convert moles of NO to L of NO. The moles of a gas at STP are converted to L by multiplying by the molar volume, 22.4 L per mole:

$$(2.3 \text{ mol NO})\left(\frac{22.4 \text{ L}}{\text{mol}}\right) = 52 \text{ L NO}$$

8. 1.5×10^2 L (2 significant figures)

We must first calculate how many moles 15 g of H_2 represents and then solve the equation for "volume".

$$(15 \text{ g})\left(\frac{1 \text{ mol } H_2}{2.0 \text{ g}}\right) = 7.5 \text{ mol } H_2$$

$PV = nRT$ or $V = \dfrac{nRT}{P}$

R = 0.0821 L • atm/mol • K

$$V = \frac{(7.5 \text{ mol})\left(0.0821 \dfrac{\text{L·atm}}{\text{mol·K}}\right)(298 \text{ K})}{1.2 \text{ atm}}$$
$$= 1.5 \times 10^2 \text{ L}$$

9.   134 L

This is a volume-volume calculation and is an application of Avogadro's Law. Also, we need to consider whether there is a limiting reactant. First, determine how many moles of NO(g) can be produced from each reactant.

$$(4.00 \text{ mol N}_2)\left(\frac{2 \text{ mol NO}}{1 \text{ mol N}_2}\right) = 8.00 \text{ mol NO}$$

$$(3.00 \text{ mol O}_2)\left(\frac{2 \text{ mol NO}}{1 \text{ mol O}_2}\right) = 6.00 \text{ mol NO}$$

Therefore, O_2 is the limiting reactant and 6.00 mol NO will be produced in the reaction with 1.00 mol of N_2 left over.

The volume of 6.00 mol of NO at STP will be

$$(6.00 \text{ mol})\left(22.4 \frac{\text{L}}{\text{mol}}\right) = 134 \text{ L}$$

10.   Almond odor

The molar mass of arsine is 77.94 g/mol and that of cyanogen is 52.04 g/mol. since the rate of effusion is inversely proportional to the square roots of their molar masses, the hero will smell almonds first.

11. 1.80×10^2 moles O_2 and 5.76×10^3 g O_2

We need to use the ideal gas equation to solve the problem. The data tabulated looks like this:

$$P = (1500 \text{ psi}) = (1500 \text{ psi})\left(0.06805 \frac{\text{atm}}{\text{psi}}\right) = 1.0 \times 10^2 \text{ atm}$$

$$V = 43.8 \text{ L}$$

$$R = 0.0821 \frac{\text{L·atm}}{\text{mol·K}}$$

$$T = 72°\text{F} = 22 °\text{C} = 295 \text{ K}$$

$$PV = nRT$$

$$n = \frac{PV}{RT} = \frac{(1.0 \times 10^2 \text{ atm})(43.8 \text{ L})}{\left(0.0821 \frac{\text{L·atm}}{\text{mol·K}}\right)(295 \text{ K})} = 1.80 \times 10^2 \text{ mol O}_2$$

The mass of oxygen equals

$$(1.80 \times 10^2 \text{ mol } O_2)\left(32.00 \frac{\text{g}}{\text{mol}}\right) = 5.76 \times 10^3 \text{g}$$

12. 77.7 g/mol

First, we need to convert torr into atm and °C into K.

$$(740 \text{ torr})\left(\frac{1 \text{ atm}}{760 \text{ torr}}\right) = 0.974 \text{ atm}$$
$$21°C + 273 = 294 \text{ K}$$

The ideal gas equation is: $PV = nRT$

Rearranging and using grams/molar mass $= n$, we have

$$\text{molar mass (M)} = \frac{g \, RT}{PV}$$

Substituting into the equation:

$$M = \frac{(3.48g)(0.0821 \, \text{L·atm})(294 \, \text{K})}{(0.974 \, \text{atm})(1.11 \, \text{L})(\text{mol·K})}$$
$$= 77.7 \text{ g/mol}$$

This value varies somewhat from the accepted value of 77.94 g/mol because of rounding off several of the given values.

CHAPTER THIRTEEN

Properties of Liquids

SELECTED CONCEPTS REVISITED

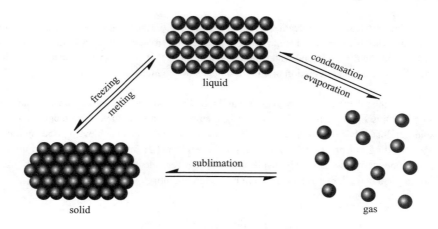

The molecules in solids have low average kinetic energy. As the temperature is raised the molecules gain kinetic energy raising the average kinetic energy of the molecules in the sample. If the molecules gain enough energy, they eventually break enough of the intermolecular attractions to become a liquid and if all the attractions are overcome, a gas.

The vapor pressure of a liquid is the pressure exerted by the vapor above a liquid when the rate of evaporation is equal to the rate of condensation. Using water as an example, suppose you have a sample of water in a closed container. Some of the water molecules may possess enough kinetic energy to break free of the other molecules and achieve the gaseous state (that is, some molecules will become water vapor). The water vapor (a gas) will exert pressure on the walls of the container. Eventually some of the molecules of the water vapor may lose kinetic energy and condense to the liquid state. At some point in time this process will achieve equilibrium, that is the rate of the two processes (vaporization and condensation) will be equal. The vapor pressure is the pressure exerted by the vapor when a liquid is in equilibrium with its vapor. Because the average kinetic energy of liquid is dependent on its temperature, the vapor pressure of a liquid is also temperature dependent.

Liquids with a higher vapor pressure evaporate faster and the liquid is considered to be more volatile. Liquids with higher vapor pressures tend to have less intermolecular attractions between liquid molecules and therefore do not need as much energy to overcome these attractions. Note that you need not memorize this; it is better to understand vapor pressure and you can then rationalize the trend. If you have two liquids under similar conditions and assume there is no vapor initially present, then when both have reached the liquid-gas equilibrium for the same temperature, the only way one liquid can have a higher vapor pressure is if more molecules are in the gaseous state at equilibrium. So higher vapor pressure means more gas molecules present when rate of evaporation equal rate of condensation therefore it is probably easier to vaporize that liquid so there are weaker intermolecular attractions in the liquid state.

The boiling point of a liquid is the temperature at which its vapor pressure is equal to the external (atmospheric) pressure. Therefore changing the external pressure can alter the actual boiling point of a liquid. If the external pressure is raised, that means the boiling point of the liquid is also raised. Here is a possible rationalization:

Higher external pressure means the liquid needs to have a higher vapor pressure before it will boil. A higher vapor pressure means more molecules need to have escaped from the liquid to get into the gas state. To get more molecules into the gas state, the liquid needs to have a higher average kinetic energy. Higher temperatures cause higher kinetic energies of molecules. So higher external pressure causes a higher boiling point.
Alternatively, low external pressure means a lower vapor pressure is necessary therefore less molecules in the vapor state and so the liquid can boil at a lower temperature.

Water, H_2O, is a polar covalent bent molecule, with the O having a $\delta-$ and each H having a $\delta+$ charge. (δ means partial.) Liquid water exhibits an intermolecular attraction called a **hydrogen bond**. A hydrogen bond is an electrostatic attraction between the $\delta-$ O of one water molecule and the $\delta+$ H on another molecule. A hydrogen bond is simply such a good dipole-dipole interaction that it warrants its own name.

A hydrogen bond is formed between a really good $\delta-$ and a really good $\delta+$. How do we know what constitutes a really good $\delta-$ or $\delta+$? Only the three most electronegative elements are able to have a good enough $\delta-$ to participate in hydrogen bonding; thus only $\delta-$ which reside on F, O, and/or N. Only H can achieve a good enough $\delta+$ to participate in hydrogen bonding but not all H's. To get a good $\delta+$ H, that H must be bonded to an atom which really shares the electrons unequally, that is the H must be attached to a very electronegative atom. So only H's attached to an F, O or N have a good enough $\delta+$ to participate in hydrogen bonding.

COMMON PITFALLS TO AVOID

Do not confuse boiling point with vapor pressure. The normal boiling point is the temperature at which the vapor pressure is equal to the atmospheric pressure (1 atm) above the liquid. The vapor pressure is the pressure exerted by the vapor above the liquid when the liquid and its vapor are at equilibrium. The vapor pressure is temperature dependent.

Remember that it takes energy to actually carry out a phase change. That is, in going from a solid to a liquid, there will be some time when the solid is absorbing heat but there is no visible change in temperature. The heat being absorbed during this time is being used to break some of the intermolecular attractions and not into raising the temperature. A similar situation occurs in going from a liquid to a gas. So if you are trying to determine how much heat is absorbed when ice at $-10\,°C$ to water at $50\,°C$ do not forget to include the heat absorbed during the phase change when there is no temperature change.

A hydrogen bond is not a covalent bond! Only H's attached to F, O, or N can participate in hydrogen bonding but please remember that those covalent bonds are not hydrogen bonds.

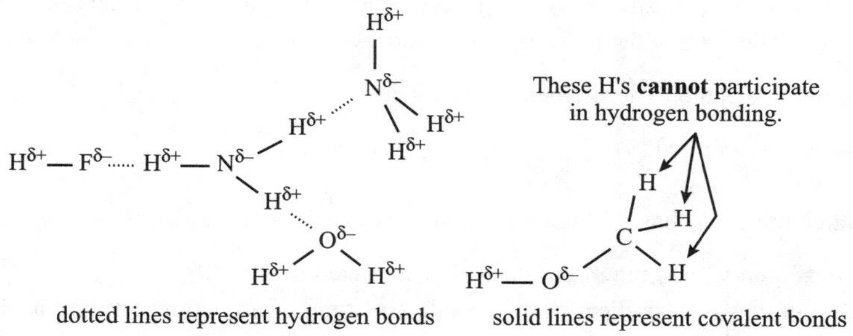

$H^{\delta+}$

$N^{\delta-}$

$H^{\delta+}$ — $F^{\delta-}$ ...... $H^{\delta+}$ — $N^{\delta-}$

$H^{\delta+}$

$H^{\delta+}$

$H^{\delta+}$

$H^{\delta+}$

$O^{\delta-}$

$H^{\delta+}$ $H^{\delta+}$

These H's **cannot** participate in hydrogen bonding.

H

C — H

H

$H^{\delta+}$ — $O^{\delta-}$

dotted lines represent hydrogen bonds   solid lines represent covalent bonds

(Note, it is unlikely you will have a flask containing these molecules together. They are shown here to illustrate hydrogen bonding.)

When water boils, H_2O molecules remain as H_2O molecules. The hydrogen bonds between H_2O molecules are broken but not the bonds in the H_2O molecule itself.

SELF-EVALUATION SECTION

1. Many of the important characteristics of water are related to its physical properties. For example, the amount of heat required to change 1 gram of a solid into a liquid, called the (1) _____ , is an unusually large amount for water, as is the amount of heat required to change 1 gram of liquid at its normal boiling point into a gas called the (2) _____ . The temperature at which ice begins to change into the liquid state is called the (3) _____ and the temperature at which the vapor pressure of water equals the atmospheric pressure of 1 atm is called the (4) _____ . Water reacts with numerous compounds to form useful products. For example, certain metallic oxides react with water to form bases and are known as (5) _____ , while nonmetallic oxides form acids with water and are known as (6) _____ . Calcium oxide (CaO) is a basic anhydride which is used in the cement industry, and sulfur trioxide (SO_3) is the acid anhydride of sulfuric acid (H_2SO_4). Other compounds contain water molecules as part of their crystalline structure and are called (7) _____ . One often sees dramatic color changes when the water of hydration is removed from a hydrated salt.

2. Water, which is the most common chemical substance around us, has been the subject of many experiments over the years. It is a simple molecule, H_2O, and yet its properties suggest that water is a very large molecule. We want to analyze how the structure of water influences such properties as boiling point and melting point.

    Fill in the blank space or circle the appropriate response.

    A single water molecule consists of (1) _____ H atom(s) and (2) _____ O atom(s). The oxygen atom is the middle atom joined to each H atom by a(n) (3) <u>ionic/covalent</u> bond. The molecule is (4) <u>straight end to end/bent in the middle</u> with a bond angle of 105° between the two covalent bonds.

Oxygen is a very (5) <u>electropositive/electronegative</u> element and, as a result, the two covalent OH bonds are (6) <u>nonpolar/polar</u>. The bend in the molecule in conjunction with the polar covalent bonds make water a

(7) <u>nonpolar/polar</u> molecule. The oxygen atom carries a partial (8) _____ charge, and each

hydrogen atom carries a partial (9) _____ charge.

Since each water molecule has the same unequal charge distribution, there will be an attraction between

molecules. The oxygen side of the molecule will be attracted to the (10) _____ atom of another water molecule through a weak electrostatic bond. This type of attraction between an H atom and a highly

electronegative atom is called a (11) _____ bond. The weak ionic association between water molecules produces the effect of water behaving as a large molecule. When we examine the physical properties of water, it is clear that water does not fit the expected pattern. The melting point and normal boiling point are (12) <u>higher/lower</u> than expected, as are the heat of fusion and heat of vaporization. Water acts though it were a large bulky molecule rather than a small one with a molar mass of 18.02 g/mol. The bent structure and resulting polarity have a marked influence on the physical properties of water.

3. Fill in the blank space.

Natural fresh waters are usually not pure enough to drink and therefore must be treated. Removal of large

objects is accomplished by (1) _____ whereas fine particles are removed by (2)

_____ and (3)_____ . The last step (4) _____ kills bacteria. If it con-

tains dissolved magnesium and calcium salts, the water is said to be (5) _____ . Three techniques

used to soften hard water are (6) _____ , (7) _____ , and (8) _____ .
The process that uses zeolite to soften hard water is a type of (9) _____ technique.

4. Fill in the blank space or circle the appropriate response.
All substances in the liquid state are in the process of vaporizing. Some chemicals, such as acetone and ether, evaporate quickly, whereas others, such as mercury, do so slowly. The process of molecules going from the

liquid state to the gas state is called (1) _____ . In any sample of a liquid, (2) <u>repulsive/attractive</u> forces exist that must be overcome before a molecule can escape the liquid state. Even though the temperature of the liquid is uniform, not all of the molecules possess the same (3) <u>kinetic/potential</u> energy. Since the masses are constant, this suggests that the velocities of the molecules are (4) <u>the same/different</u>. Therefore, molecules at the surface of the liquid, which are moving faster than their neighbors, are able to overcome the attractive forces and escape to the gas state. After these molecules with greater kinetic energy leave the liquid state, the average kinetic energy of the remaining molecules in the liquid state is (5) <u>lowered/raised</u>. There-fore, the temperature (6) <u>raises/ drops</u>, and we find that vaporization is a (7) <u>warming/cooling</u> process, which we know to be true from everyday experience. If the temperature drops as the average kinetic energy of the system goes down, then where does the heat energy come from to vaporize all the water from an open dish?

(8) _____

5. Fill in the blanks.

A liquid boils when its (1)_____ _____ is equal to the (2) _____ _____. The latter is usually considered to be (3) _____ torr, but can vary with the weather and elevation difference from that of sea level; it (4) _____ with increased elevation.

6. At what temperature will ethyl ether boil when subjected to an external pressure of 400 torr? Refer to Figure 13.6 in your text.

7. On the following heating curve, a solid substance at point A is heated until it reaches point E. Identify the various stages along the curve as requested.

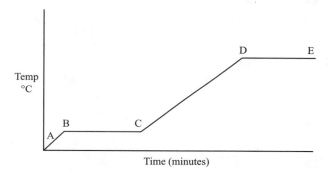

Identification of Process of Condition of State

(1)   Point B   _____

(2)   Line BC   _____

(3)   Line CD   _____

(4)   Point D   _____

(5)   Line DE   _____

8. Complete and balance the following reactions involving water as a reactant.

(1) $K(s) + H_2O(l) \rightarrow H_2(g) +$
(2) $Al(s) + H_2O(steam) \rightarrow \qquad + Al_2O_3(s)$
(3) $Fe(s) + H_2O(steam) \rightarrow H_2(g) +$
(4) $Cl_2(g) + H_2O(l) \rightarrow \qquad + HOCl(aq)$

9. Identify each of the salt formulas as anhydrous or hydrates. Name them.

(1) $CaSO_4$ _____

(2) $CoCl_2 \cdot 6\ H_2O$ _____

(3) $NaC_2H_3O_2 \cdot 3\ H_2O$ _____

(4) K_2S _____

(5) Na_3PO_4 _____

10. Identify the compounds listed as basic anhydrides or acidic anhydrides and write their reactions with water.

    (1) CaO   (3) N_2O_5
    (2) SO_3 (4) Na_2O

    Write your answers below.

11. Write the formulas for the anhydrides of the following.

    (1) H_2SO_3, $HClO_4$, H_2CO_3
    (2) KOH, $Ba(OH)_2$, $Mg(OH)_2$

    Write your answers below.

Challenge Problems

12. An ice cube at 0°C has a mass of 7.25 g.
    (1) In order for the ice cube to be heated up enough to vaporize, which stages of the heating curve must it pass through?
    (2) Use your answer to help you calculate the heat (in joules) needed to convert the ice cube at 0°C to steam at 100°C.

    The heat of fusion is 335 J/g. The heat of vaporization is 2.26 kJ/g. The specific heat is 4.184 J/g°C.

    Do your calculations here.

13. How many calories of heat energy are required to vaporize two 35.0 g pieces of solid ethanol, C_2H_5OH, at $-112°C$ given the following information.

Ethanol, C_2H_5OH melting point $-112°C$
boiling point $78.4°C$
Heat of fusion 24.9 cal/g
Heat of vaporization 204.3 cal/g

specific heat, C_p, 2.49 $\dfrac{\text{joule}}{\text{g °C}}$

calories = (joules)$\left(0.239 \dfrac{\text{cal}}{\text{joule}}\right)$

Do your calculations here.

RECAP SECTION

Chapter 13 introduces many new terms you should be familiar with. We discussed a chemical compound that we are in contact with every day. Much of the discussion centered around the application of previously learned concepts. The importance of electronegativity in explaining the physical properties of water was good exercise for you. Since some of the topics in the chapter were not covered in the study guide, you may want to go back over the sections on the chemistry of water on your own. Now that you have learned some things about water by itself, let us begin to use water as we do in the laboratory to prepare chemical solutions. In the next unit, we discuss the principal use of water in chemistry – as a solvent for reagents.

ANSWERS TO QUESTIONS

1. (1) heat of fusion   (2) heat of vaporization   (3) melting point
    (4) normal boiling point   (5) basic anhydrides   (6) acid anhydrides
    (7) hydrates

2. (1) 2   (2) 1   (3) covalent
    (4) bent in the middle   (5) electronegative   (6) polar
    (7) polar   (8) negative   (9) positive
    (10) hydrogen   (11) hydrogen   (12) higher

3. (1) screening   (2) flocculation   (3) sedimentation
    (4) disinfection   (5) hard
    (6), (7), (8) precipitation, distillation, ion exchange, or demineralization (9) ion exchange

4.	(1)	vaporization	(2)	attractive	(3)	kinetic
	(4)	different	(5)	lowered	(6)	drops
	(7)	cooling	(8)	the surroundings

5.	(1) vapor pressure	(2) atmospheric pressure	(3) 760	(4) decreases

6.	Using the vapour-pressure-temperature curve in Figure 13.6, the temperature corresponding to 400 torr is approximately 17 °C.

7.	(1)	melting point	(2)	solid in equilibrium with liquid
	(3)	all-liquid state, temperature	(4)	boiling point
		begins to rise
	(5)	boiling liquid in equilibrium
		with gas

8.	(1)	$2\,K(s) + 2\,H_2O(l) \rightarrow H_2(g) + 2\,KOH(aq)$
	(2)	$2\,Al(s) + 3\,H_2O(steam) \rightarrow 3\,H_2(g) + Al_2O_3(s)$
	(3)	$3\,Fe(s) + 4\,H_2O(steam) \rightarrow 4\,H_2(g) + Fe_3O_4(s)$
	(4)	$Cl_2(g) + H_2O(l) \rightarrow HCl(aq) + HOCl(aq)$

9.	(1)	anhydrous	(2)	hydrate	(3)	hydrate
		calcium sulfate		cobalt (II) chloride		sodium acetate
				hexahydrate		trihydrate

	(4)	anhydrous	(5)	anhydrous
		potassium sulfide		sodium phosphate

10.	(1)	CaO – basic anhydride
		$CaO + H_2O \rightarrow Ca(OH)_2$

	(2)	SO_3 – acidic anhydride
		$SO_3 + H_2O \rightarrow H_2SO_4$

	(3)	N_2O_5 – acidic anhydride
		$N_2O_5 + H_2O \rightarrow 2\,HNO_3$

	(4)	Na_2O – basic anhydride
		$Na_2O + H_2O \rightarrow 2\,NaOH$

11.	(1)	SO_2	Cl_2O_7	CO_2
	(2)	K_2O	BaO	MgO

12.	26.4 kJ
	(1)	The ice cube passes through 3 stages …
			melting the ice:	ice at 0°C → water at 0°C
			heating the water:	water at 0°C → water at 100°C
			vaporizing the water:	water at 100°C → steam at 100°C

To melt the ice cube requires 335 J for each gram. Therefore,

$$\left(335\frac{J}{g}\right)7.25\ g = 2430\ J$$

We now have water at 0°C and will need 4.184 J per gram of water per degree to raise the water temperature to boiling. Therefore,

$$(7.25\ g)(100°C)\left(4.184\ \frac{J}{g°C}\right) = 3030\ J$$

Notice how the units cancel to give us units of heat energy – joules. Now we must use 2.26 kJ for each gram of hot water to convert it into steam.

$$(7.25\ g)\left(2.26\ \frac{kJ}{g}\right) = 16.4\ kJ$$

The last step is to add up the individual values after converting to kJ.

$$2.43\ kJ + 3.03\ kJ + 16.4\ kJ = 21.9\ kJ$$

(If you waited until the end of the calculation to round off, the answer is 21.865)

13. 1.71×10^4 calories

The first step is to melt the two 35.0 g pieces of solid ethanol at -112°C.

$$(70.0\ g)(24.9\ cal/g) = 1.74 \times 10^3\ cal\ (3\ significant\ figures)$$

The temperature must be raised to the boiling point, 78.4°C. The specific heat is given in joules and must be changed to calories.

$$(2.49\ joule/°C\ g)(0.239\ cal/joule)(70.0\ g)(temperature\ change)$$
$$(2.49\ joule/°C\ g)(0.239\ cal/joule)(70.0\ g)(190.4°C) = 7.43 \times 10^3\ cal$$

To vaporize the 50.0 g of liquid at the boiling point requires 204.3 cal/g

$$(70.0\ g)(204.3\ cal/g) = 1.43 \times 10^4\ cal$$

Adding up the 3 values gives the final answer

$$(1740\ cal) + (7930\ cal) + (14,300\ cal) = 23970 = 2.40 \times 10^4\ kcal$$

Solutions

SELECTED CONCEPTS REVISITED

True liquid solutions are homogeneous mixtures and often visibly transparent. The solute is molecular or ionic in size, does not settle out, and can often be easily separated from the solvent by physical means.

Two nonpolar substances easily intermingle to form a solution. The dissolution of ionic compounds in water is somewhat more complicated. Water is very polar and can attract positively and negatively charged ions. These attractions are generally strong enough to allow the ions to move away from each other with the net result being that the ions are distributed in the solution. Each ion is surrounded by water molecules whose opposite pole is facing inwards (that is, the hydrogens of water face in towards a negative charge while the oxygen of water face inwards to a positive charge).

The effect of temperature on the solubility of a solid in a liquid is somewhat unpredictable. For some solids, their solubility decreases but for most solids the solubility increases. The temperature effect on the solubility of gases is more predictable. In general, as the temperature of the solution increases, the solubility of the gas in the solution decreases.

A change in pressure tends to significantly affect the solubility of a gas in a liquid. The solubility of the gas tends to rise proportionately to an increase in pressure.

Qualitative descriptions of aqueous solutions include the following terms:
Dilute – only a small amount of solute is present *relative* to the volume of the solution.
Concentrated – a large amount of solute is present *relative* to the volume of the solution.
Unsaturated – the solution is capable of dissolving more solute.
Saturated – the maximum amount of solute relative to the volume of the solution has been dissolved. Any excess solute added will not dissolve. The rate of solute dissolving to the rate of solute crystallizing out of solution is equal (at equilibrium).
Supersaturated – more solute than can normally be dissolved in solution has dissolved. Note that this type of solution must be prepared carefully and the excess solute is easily crystallized out (often triggered by something as simple as moving the container.)

Quantitative descriptions of the concentration of aqueous solutions include the following units: Mass percent, ppm, mass/volume %, volume %, molality, and perhaps the most commonly used term, molarity.

The concentration of a solution is most often described using units of molarity (M).
M = moles solute/liters of solution.

In a dilution, the volume of the solvent is altered, therefore the volume of the solution also changes. Since the moles of solute remain constant but the volume of the solution changes, the molarity (or concentration) of the solution also changes.

Colligative properties are the properties of a solution that depend on the number of particles in solution. For example, consider the effect of an increased number of nonvolatile particles on the boiling point. The more nonvolatile particles in solution, the lower the vapor pressure which in turn means that more heat needs to be applied to the solution to get more molecules into the vapor phase before the vapor pressure equals the atmospheric pressure which leads to a higher boiling point.

COMMON PITFALLS TO AVOID

Not all liquids are solutions, not all solutions are liquids (although you will deal mainly with liquid aqueous solutions) and not all mixtures are solutions.

Saturated is not the same as concentrated. A solution could be saturated and still be dilute. For example, if a solute is not very soluble in water, then even when as much solute as possible has been dissolved, relative to the volume of water present the amount of solute is small and therefore even though the solution is saturated, it may still be considered dilute.

Molarity and molality may sound similar but these two units of concentration are not the same. Molarity is generally used to describe most aqueous solution and has units of mol/L. Molality is more commonly used in freezing point depression or boiling point elevation calculations and has units of mol/kg.

Realize that $M_1L_1 = M_2L_2$ is a shortcut for a dilution problem. You should only use shortcuts that you understand. Molarity × volume (in L) gives moles of solute. Since the moles of solute do not change during a dilution, the molarity × volume before dilution is equal to the molarity × volume after the dilution. Gaining a good understanding of the concept behind $M_1L_1 = M_2L_2$ will help you use the formula only at the appropriate times.

Use caution when using or interpreting equation symbols. For example, M represents molarity, m can represent mass or molality, and mol (not m) represents moles. It should be noted that molality is often an italicized m in texts but in hand-written formula, it is near impossible to distinguish from the mass, m. Be sure to use the appropriate quantity for your calculations. K's (and k's) are often also used in multiple situations.

SELF-EVALUATION SECTION

1.   In the following statements, identify the (a) solution, (b) solute, and (c) solvent.
     (1)   Household bleach is a 5% solution of sodium hypochlorite (NaClO) in water.

          a. _____   b. _____   c. _____

     (2)   A 0.1 M iodine solution was prepared by dissolving crystals of iodine in carbon tetrachloride (CCl_4).

          a. _____   b. _____   c. _____

(3)   Air is composed primarily of two gases – oxygen and nitrogen. Air is approximately 79% nitrogen and 21% oxygen.

    a. _____   b. _____   c. _____

(4)   Nickel coins in the United States are made from a nickel-copper alloy that is 75% copper and 25% nickel.

    a. _____   b. _____   c. _____

(5)   In order for certain species of fish to thrive in lakes and streams, the dissolved oxygen content of the water has to be 0.0005% or greater.

    a. _____   b. _____   c. _____

2.   Solubility of salts in water.

All nitrates are (1) _____ in water.

All chlorides, bromides, and iodides are (2) _____ in water except those of silver, mercury (I), and lead (II).

All carbohydrates and phosphates are (3) _____ in water except those of sodium, potassium, and ammonium.

All sulfides are (4) _____ in water except those of ammonium sulfide and Group 1A and Group 2A metals.

Place an s (for soluble) or i (for insoluble) next to each of these compounds.

(5)	Na_2S	_____	(15)	$(Al(NO_3)_3 \cdot 9\ H_2O$	_____
(6)	Na_2CO_3	_____	(16)	K_3PO_4	_____
(7)	BaI_2	_____	(17)	$NH_4)_2CO_3$	_____
(8)	NH_4NO_3	_____	(18)	KCl	_____
(9)	$NiCO_3$	_____	(19)	$Mg_3(PO_4)_2$	_____
(10)	$LiI \cdot 3\ H_2O$	_____	(20)	$PbCl_2$	_____
(11)	$Sn(NO_3)_4$	_____	(21)	$HgBr$	_____
(12)	Fe_2S_3	_____	(22)	$ZrCl_4$	_____
(13)	Fe_2S_3	_____	(23)	Ag_2S	_____
(14)	AgI	_____	(24)	$CaBr_2$	_____

3.   Solubilities of salts in g/100 g H_2O.

$CaBr_2$	125 g at 0°C	$(NH_4)_2CO_3$	100 g at 15°C
NH_4NO_3	118 g at 0°C	$NaC_2H_3O_2$	119 g at 0°C
Na_2S	15.4 g at 10°C	$LiI \cdot 3\ H_2O$	151 g at 0°C
$Na_2S_2O_3$	50 g at 20°C	KCl	34.7 g at 20°C
$MgSO_4$	26 g at 0°C	Na_2CO_3	7.1 g at 0°C
$Al(NO_3)_2$	63.7 g at 25°C	K_3PO_4	90 g at 20°C

Identify the following solutions as saturated, unsaturated, or supersaturated. (All solutes are in 100 g H_2O.)

(1) 5 g of $CaBr_2$ at 0°C

(2) 100 g of NH_4NO_3 at 0°C

(3) 15.4 g of Na_2S at 10°C

(4) 55 g of $Na_2S_2O_3$ at 20°C

(5) 4 g of $MgSO_4$ at 0°C

(6) 51 g of $Al(NO_3)_3$ at 25°C

(7) 75 g of $(NH_4)_2CO_3$ at 15°C

(8) 125 g of $NaC_2H_3O_2$ at 0°C

(9) 38 g of $LiI \cdot 3H_2O$ at 0°C

(10) 34.7 g of KCl at 20°C

(11) 6.5 g of Na_2CO_3 at 0°C

(12) 12 g of K_3PO_4 at 20°C

4. Fill in the blank space or circle the appropriate response.

Water molecules, which are very (1) polar/nonpolar, are (2) attracted/repulsed by other polar or ionic molecules or ions. Water molecules weaken the ionic forces that hold ions such as Na^+ and Cl^- together. Then the ions are pulled apart by the interaction with water as the water molecules surround the ions. The charged ions such as Na^+ and Cl^- then diffuse slowly away from the mass of undissolved salt as hydrated ions.

With most solid chemicals, an increase in temperature means an (3) _____ in solubility. However, a gaseous chemical always (4) _____ in solubility as the temperature increases. A temperature increase means that the kinetic energy of the gas molecules (5) _____ and their ability to associate with the solvent molecules (6) _____ .

Pressure changes don't affect the solubility of solids greatly but gases show marked changes. The solubility of a gas is (7) inversely/directly proportional to the pressure of the gas above the liquid. Double the pressure means the solubility will (8) _____ . Carbonated beverages are a good example.

5. The solubility of salts can vary considerably with temperature. Although the solubility of most substances increase with increasing temperature, some decrease. Using Figure 14.4 in your text, list the following solutes in order of increasing solubility.

KCl, $CuSO_4$, HCl, Li_2SO_4, KNO_3 (all at 1 atm)

(1) At 10°C _____ , _____ , _____ , _____ , _____ (most soluble)

(2) At 50°C _____ , _____ , _____ , _____ , _____ (most soluble)

(3) If 20 g of each solute above was mixed into 50 g of water, in which flasks would you expect to see undissolved solid?

6. List the four factors that influence the rate or speed at which a solid solute dissolves.

7. (1) How would you prepare 1500 g of a 2.000% sucrose (sugar) mass percent solution to be used for intravenous feeding in a hospital?

   Do your calculations here.

   (2) What masses of sodium chloride (NaCl) and water are needed to make 525.0 g of 0.1000% salt solution?

   Do your calculations here.

8. A 1.5 L bottle of cooking sherry contains 18% alcohol by volume.

   (1) How many mL of alcohol are in the bottle?

   Do your calculations here.

   (2) How many mL of alcohol are in a 34 mL (1/8 cup) measure of sherry?

   Do your calculations here.

9. (1) A 250. mL solution contains 3.8 moles of H2SO4. What is the molarity of the solution?

    Do your calculations here.

    (2) If 2.6 moles of KCl is used to make a 4.0 L batch of solution, calculate the molarity of the solution.

    Do your calculations here.

10. Calculate how many grams of chemical would be required to prepare the following solutions.

Atomic Masses, amu	
Na = 22.99	F = 19.00
S = 32.07	Ca = 40.08
O = 16.00	C = 12.01
N = 14.01	H = 1.008

(1) 600. mL of 3.00 M NaOH
(2) 4.00 L of 1.00 M $(NH_4)_2SO_4$
(3) 420 mL of 0.70 M $CaSO_4$
(4) 250 mL of 0.94 M $Ca(NO_3)_2$
(5) 0.50 L of 0.10 M NaF

Use the relationship

$$M = \frac{g \text{ of solute}}{\text{molar mass solute} \times L \text{ of solution}}$$

or solve by dimensional analysis.

Do your calculations here.

11. More experience in working with molarities of solutions.

   (1)   What volume of 1.25 M $K_2Cr_2O_7$ can be prepared from 130. g of $K_2Cr_2O_7$?

Atomic Masses, amu

K = 39.10

Cr = 52.00

O = 16.00

   Do your calculations here.

   (2)   What volume of 0.100 M $NaHCO_3$ can be prepared from 90.0 g of $NaHCO_3$?

Atomic Masses, amu

Na = 22.99

H = 1.008

O = 16.00

C = 12.01

   Do your calculations here.

   (3)   Calculate the number of moles of hydrochloric acid in 4.0 L of 0.333 M HCl.

   Do your calculations here.

(4)   Calculate the number of moles of solute contained in the following solutions:

250 mL of 0.22 M $K_2Cr_2O_7$
1500 mL of 1.4 M Na_2CO_3
3.15 L of 0.75 M $HClO_4$
857 mL of 0.66 M $CuSO_4 \cdot 5\ H_2O$
50. mL of 0.25 M NaS_2O_3

Do your calculations here.

(5)   Using the relationship for dilutions, calculate how much (mL) concentrated reagent is necessary for the following solutions.

$$(\text{volume}_1)(M_1) = (\text{volume}_2)(M_2)$$

15 M NH_4OH to prepare 2.5 L of 5.0 M NH_4OH
14.6 M H_3PO_4 to prepare 150. mL of 0.30 M H_3PO_4
12 M HCl to prepare 500. mL of 0.25 M HCl
18 M H_2SO_4 to prepare 250. mL of 1.25 M H_2SO_4

Do your calculations here.

12. Water from the Great Salt Lake can have a salt concentration as high 3.42 M (expressed as NaCl). High mountain spring water has a very low salt concentration, 0.00171 M. Potable, or drinkable, water has a maximum recommended level of 0.0171 M or 10 times that of spring water. What is the maximum volume of drinkable water that can be prepared from 500 mL of Great Salt Lake water using spring water for dilution? Assume the salt concentration from the spring water is negligible.

Do your calculations here.

13. Lead ion can be precipitated out of solution according to the following reaction.

$$Pb^{2+}(aq) + Na_2CrO_4(aq) \rightarrow PbCrO_4(s) + 2\ Na^+(aq)$$

What mass of Na_2CrO_4 should be added to 10.0 L of solution that contains 2.78 g/L of Pb^{2+} ion?

Do your calculations here.

14. Colligative properties of solutions depend only on the (1) _____ of solute particles present. Therefore, 1 mole of sugar and 1 mole of alcohol, neither of which is ionic, will depress the freezing point or elevate the (2) _____ of a fixed amount of water by an (3) equal/unequal amount. These properties are usually expressed on the basis of a fixed amount of solvent, which is (4) _____ , whether water or some other solvent. Other colligative properties include (5) _____ , in addition to freezing point depression and boiling point elevation. A mole of an ionic substance such as NaCl will lower the

freezing point of a solution (6) _____ as much as a nonionic material since (7) _____ ions are produced for each mole of NaCl. A mole of $CaCl_2$ will lower the freezing point of water (8) _____ times as much as a mole of sugar or urea. For un-ionized and non-volatile substances, molecular masses can be determined from either of two colligative properties, namely, (9) _____ and (10) _____ .

15. (1) How much ethylene glycol, $C_2H_4(OH)_2$, per kilogram of water is needed to lower the freezing from $0\,°C$ to $-20\,°C$?

The value for K_f is $1.86°C/mol/kg$. The equation is:

$$\Delta t_f = (K_f)\left(\frac{\text{grams solute}}{\text{molar mass solute}}\right)\left(\frac{1}{\text{kg solvent}}\right)$$

Rearranging the equation to solve for grams of solute gives us

$$\text{g solute} = \frac{(\Delta t_f)(\text{molar mass})(\text{kg solvent})}{K_f}$$

First determine the molar mass of ethylene glycol and then make the necessary substitutions in order to solve the equation.

Do your calculations here.

(2)   A sample of an organic compound having a mass of 1.50 g lowered the freezing point of 20.0 g of benzene by 2.75 °C. The K_f of benzene is 5.1 °C/mol/kg. Calculate the molar mass of the compound. The equation is:

$$\Delta t_f = (K_f)\left(\frac{\text{grams solute}}{\text{molar mass solute}}\right)\left(\frac{1}{\text{kg solvent}}\right)$$

Rearranging the equation to solve for molar mass gives us

$$\text{molar mass} = \frac{(K_f)(\text{g solute})}{(\Delta t_f)(\text{kg solvent})}$$

Make the necessary substitutions and calculate the molar mass of the unknown compound.

Do your calculations here.

Challenge Problems

16.   Cadmium ion, which is a toxic metallic ion, can be precipitated from solution according to the following reaction.

$$Cd^{2+}(aq) + K_3PO_4(aq) \rightarrow Cd_3(PO)_4(s) + K^+(aq)$$

Balance the equation and, assuming no side reactions, determine what volume of stock 5.00 M K_3PO_4 solution should be diluted to 250. mL in order to precipitate 4.10 g Cd^{2+} ion that is contained in 600. mL of a waste solution.

Do your calculations here.

17. One gram (1.00 g) of glucose ($C_6H_{12}O_6$) dissolves in 1.10 mL of H_2O. What is the concentration in terms of mass percent and molality? Human blood normally contains approximately 0.090% glucose (mass percent). What is the molality of blood glucose?

Assume 1000 g of blood is equivalent to 1000 g of solvent.

Do your calculations here.

RECAP SECTION

After completing Chapter 14, you have learned many of the skills that a bench chemist or laboratory technician uses every day. These people often work with solutions; therefore, the knowledge of solution preparation and calculations involving solutions is important to them as well as to other scientists and technicians. The matter of proper preparation of solutions is critical in any work situation that deals with chemicals, such as nursing, agriculture, and food technology. There are many good review exercises at the end of the chapter in the text. Do any problems that are assigned and then try a few more. This is a good opportunity to sharpen up your problem-solving ability.

ANSWERS TO QUESTIONS AND SOLUTIONS TO PROBLEMS

1.   (1)   bleach, sodium hypochlorite, water
      (2)   iodine solution, iodine, carbon tetrachloride
      (3)   air, oxygen, nitrogen
      (4)   coin alloy, nickel, copper
      (5)   lake and river water, oxygen, water

2. (1) soluble   (2) soluble   (3) insoluble   (4) insoluble

(5) s   (6) s   (7) s   (8) s   (9) i   (10) s   (11) s   (12) i   (13) i   (14) i

(15) s   (16) s   (17) s   (18) s   (19) i   (20) i   (21) i   (22) s   (23) i   (24) s

3. (1) unsaturated   (2) unsaturated   (3) saturated   (4) supersaturated

(5) unsaturated   (6) unsaturated   (7) unsaturated   (8) supersaturated

(9) unsaturated   (10) saturated   (11) unsaturated   (12) unsaturated

4. (1) polar   (2) attracted   (3) increase   (4) decrease

(5) increases   (6) decreases   (7) directly   (8) double

5. (1) $CuSO_4$, KNO_3, KCl, Li_2SO_4, HCl

(2) Li_2SO_4, $CuSO_4$, KCl, HCl, KNO_3

(3) 20 g solute in 50 g of water is equivalent to 40 g of solute per 100 g of water. Looking at Figure 14.4, we see that at 10°C, only HCl forms an unsaturated solution (and HCl is a gas so even if the solution were saturated you would not see solid anyway) so the saturated solutions of $CuSO_4$, KNO_3, KCl and Li_2SO_4 solutions will all contain undissolved solid. At 50°C only Li_2SO_4 and $CuSO_4$ will form saturated solutions and contain undissolved solid.

6. Four factors are particle size of a solute, stirring, temperature, solution concentration.

7. (1) 30.00 g sucrose, 1470 g H_2O

A 2.000% sucrose solution means that 2.000% of the total solution is sucrose.

$$\frac{2.000}{100}(1500 \text{ g}) = 30.00 \text{ g sucrose}$$

If 30.00 g of the total mass is sucrose, then the mass of water is

$$1500 \text{ g} - 30.00 \text{ g} = 1470 \text{ g } H_2O$$

(2) 0.5 g NaCl, 524.5 g H_2O

A 0.1000% solution means that 0.1000% of 525.0 g is salt.

$$\left(\frac{0.1000}{100}\right)(525.0 \text{ g}) = 0.5250 \text{ g NaCl}$$
$$= 0.5 \text{ g}$$

The mass of water is 525 g $-$ 0.5 g = 524.5 g H_2O

8. (1) 270 mL

You are asked to calculate a volume contained in a solution of a certain concentration; 18% of the total volume is alcohol.

$$\left(\frac{18}{100}\right)(1.5 \text{ L}) = 0.27 \text{ L} = 270 \text{ mL}$$

(2) 6.1 mL

Again, alcohol is 18% of the volume, so . . .

$$\left(\frac{18}{100}\right)(34 \text{ L}) = 6.1 \text{ mL}$$

9. (1) 15 M H_2SO_4

For any problem involving molarity and quantities of chemicals needed to prepare for various solutions, it is wise to work from the definitions.

From the problem, we have 3.8 moles of H_2SO_4 in 250. mL of solution.

$$M = \left(\frac{3.8 \text{ mol}}{250.\text{ mL}}\right)\left(\frac{1000 \text{ mL}}{L}\right) = 15 \text{ M}$$

(2) 0.65 M KCl

$$M = \frac{\text{mol of solute}}{\text{L of solution}} = \frac{2.6 \text{ mol}}{4.0 \text{ L}} = 0.65 \text{ M}$$

10. (1) 72 g NaOH

molar mass of NaOH = 22.99 + 16.00 g + 1.008 g = 40.00 g/mol

We need 600. mL of 3.00 M NaF.

$$M = \frac{g}{\text{molar mass} \times L}$$

Rearranging the equation to solve for g

$$g = M \times \text{molar mass} \times L$$

$$g \text{ of NaF} = (3.00 \text{ M})\left(40.00\frac{g}{\text{mol}}\right)\left(\frac{600.\text{ mL} \times 1 \text{ L}}{1000 \text{ mL}}\right)$$

We should use the complete units for M so that our answer has the correct units of grams

$$g \text{ of NaF} = \left(3.00\frac{\text{mol}}{L}\right)\left(40.00\frac{g}{\text{mol}}\right)\left(\frac{600.\text{ mL} \times 1 \text{ L}}{1000 \text{ mL}}\right)$$
$$= 72 \text{ g NaOH (2 significant figures)}$$

(2) 529 g $(NH_4)_2SO_4$

molar mass of $(NH_4)_2SO_4 = (2 \times 14.01 \text{ g}) + (8 \times 1.008 \text{ g}) + 32.07 \text{ g} + (4 \times 16.00 \text{ g})$

$$= 132.15\frac{g}{\text{mol}}$$

You need 4.00 L of 1.00 M NH_4NO_3

$$g = M \times \text{molar mass} \times L$$

$$= \left(1.00\frac{\text{mol}}{L}\right)\left(132.15\frac{g}{\text{mol}}\right)(4.00 \text{ L})$$
$$= 529 \text{ g } (NH_4)_2SO_4 \text{ (3 significant figures)}$$

(3)   40. g CaSO$_4$

molar mass of CaSO$_4$ = 40.08 g + 32.07 g + (4 × 16.00 g)

$$= 136.15\frac{g}{mol}$$

You are asked for 420 mL of 0.70 M CaSO$_4$

g = M × molar mass × L

$$= \left(0.70\frac{mol}{L}\right)\left(136.15\frac{g}{mol}\right)\left(\frac{420 \; mL \times 1 \; L}{1000 \; mL}\right)$$
$$= 40. \; g \; CaSO_4 \; (2 \; significant \; figures)$$

(4)   46 g Ca(NO$_3$)$_2$

molar mass of Ca(NO$_3$)$_2$ = 40.08 g + (2 × 14.01 g) + (6 × 16.00 g)

$$= 164.10\frac{g}{mol}$$

You are asked for 250 mL of 0.94 M Ca(NO$_3$)$_2$

g = M × molar mass × L

$$= \left(0.94 \; \frac{mol}{L}\right)\left(164.10\frac{g}{mol}\right)\left(\frac{250 \; mL \times 1 \; L}{1000 \; mL}\right)$$
$$= 46 \; g \; Ca(NO_3)_2 \; (2 \; Significant \; figures)$$

(5)   2.1 NaF

molar mass of NaF = 22.99 g + 19.00 g

$$= 41.99\frac{g}{mol}$$

You are asked for 0.50 L of 0.10 M NaF

g = M × molar mass × L

$$= \left(0.10\frac{mol}{L}\right)\left(41.99\frac{g}{mol}\right)(0.50 \; L)$$
$$= 2.1 \; NaF \; (2 \; significant \; figures)$$

11.   (1)   0.356 L or 356 mL

You are asked to solve the problem for a volume in units of L. With this in mind, write down the complete equation.

$$M = \frac{g}{molar \; mass \times L}$$

You now have to rearrange the equation so that "L" is in the numerator and isolated. Multiply both sides of the equation by "L".

$$(L)(M) = \frac{g \; (L)}{molar \; mass \; (L)}$$

Next, cancel the L's of the right side and divide both sides of the equation by M which gives you:

$$\frac{(L)(M)}{(M)} = \frac{g(L)}{\text{molar mass }(M)(L)}$$

$$(L) = \frac{g}{\text{molar mass }(M)}$$

Remember that M is the same as moles per liter and, when you divide by a fraction, the denominator of the fraction comes up to the numerator.

For example: $\dfrac{2}{1/2} = \dfrac{(2)(2)}{1} = 4$

In similar fashion we can write our last equation as

$$(L) = \frac{g}{\text{molar mass}\left(\dfrac{\text{moles}}{\text{liter}}\right)} = \frac{g(L)}{\text{molar mass (moles)}}$$

Determine the molar mass of $K_2Cr_2O_7$, substitute the data in the equation, cancel units and solve for L.

molar mass of $K_2Cr_2O_7$ = (2 × 39.10 g) + (2 × 52.00 g) + (7 × 16.00 g) = 294.20 g/mol

Therefore,

$$L = \frac{130.\cancel{g}(L)}{294.2\,\dfrac{\cancel{g}}{\cancel{\text{mol}}}(1.24\,\cancel{\text{mol}})} = 0.356L \text{ or } 356 \text{ mL}$$

(2)   10.7 L $NaHCO_3$

$$NaHCO_3 = 22.99 \text{ g} + 1.008 \text{ g} + 12.01 \text{ g} + (3 \times 16.00 \text{ g}) = 84.01\,\frac{g}{\text{mol}}$$

$$L = \frac{g}{\text{molar mass} \times M}$$

This equation is solved just like the one above in 10 (1)

$$= \frac{(90.0\,\cancel{g})(1\text{ L})}{\left(84.01\dfrac{\cancel{g}}{\cancel{\text{mol}}}\right)(0.100\,\cancel{\text{mol}})} = 10.7 \text{ L}$$

(3)   1.3 moles of HCl

For this type of problem, go back to the simple first relationship since you are not asked a question involving grams or molar mass. The defining equation for molarity is

$$M = \frac{\text{mol}}{L}$$

Therefore,

$$\text{moles} = L \times M$$

This is what you have been asked to find in the above problem.

$$\text{moles of HCl} = (4.0\,\cancel{L})\left(0.333\,\frac{\text{mol}}{\cancel{L}}\right) = 1.3 \text{ mol}$$

(4) $K_2Cr_2O_7 = 0.055$ mole, $Na_2CO_3 = 2.1$ moles.
      $HClO_4 = 2.4$ moles; $CuSO_4 \cdot 5\ H_2O = 0.57$ mole; $Na_2S_2O_3 = 0.013$ mole
                           moles $= L \times M$

$K_2Cr_2O_7$ moles $= \left(\dfrac{250\ \text{mL} \times 1\ \text{L}}{1000\ \text{mL}}\right)\left(0.22\ \dfrac{\text{mol}}{\text{L}}\right) = 0.055$ mol

Na_2CO_3 moles $= \left(\dfrac{1500\ \text{mL} \times 1\ \text{L}}{1000\ \text{mL}}\right)\left(1.4\ \dfrac{\text{mol}}{\text{L}}\right) = 2.1$ mol

$HClO_4$ moles $= (3.15\ \text{L})\left(0.75\ \dfrac{\text{mol}}{\text{L}}\right) = 2.4$ mol

$CuSO_4 \cdot 5H_2O$ moles $= \left(\dfrac{857\ \text{mL} \times 1\ \text{L}}{1000\ \text{mL}}\right)\left(0.66\ \dfrac{\text{mol}}{\text{L}}\right) = 0.57$ mol

$Na_2S_2O_3$ moles $= \left(\dfrac{50.\ \text{mL} \times 1\ \text{L}}{1000\ \text{mL}}\right)\left(0.25\ \dfrac{\text{mol}}{\text{L}}\right) = 0.013$ mol

(5) $NH_4OH = 830$ mL, $H_3PO_4 = 3.1$ mL, $HCl = 10.$ mL, $H_2SO_4 = 17$ mL

In each case, we must solve the relationship for the initial volume (V_1), which is the quantity of the concentrate reagent. Even though we dilute the reagent with water, the number of moles of chemical remains the same; the equation is a simple equality.

For the first problem, we dilute 15 M NH_4OH (ammonium hydroxide) to 2.5 L of 5.0 M NH_4OH.

Our equation is

$$(V_1)(M_1) = (V_2)(M_2)$$
$$V_1 = \frac{(V_2)(M_2)}{M_1}$$

Substituting in the equation

$$V_1 = \frac{(2.5\ \text{L})(5.0\ \text{M})}{(15\ \text{M})}$$

$V_1 = 0.83$ L or 8.3×10^2 mL of 15 M NH_4OH diluted to 2.5 L with water

H_3PO_4 You have to change 150 mL into L before substituting in the equation.

$$V_1 = \frac{(0.150\ \text{L})(0.30\ \text{M})}{(14.6\ \text{M})}$$

$V_1 = 0.0031$ L $= 3.1$ mL

HCl You have to change 500. mL into L first

$$V_1 = \frac{(0.500\ \text{L})(0.25\ \text{M})}{(12\ \text{M})}$$

$V_1 = 0.010$ L $= 10.$ mL

H_2SO_4 You have to change 500. mL into L first.

$$V_1 = \frac{(0.250 \text{ L})(1.25 \text{ M})}{(18 \text{ M})}$$

$$V_1 = 0.017 \text{ L} = 17 \text{ mL}$$

12.   100. L

This is a dilution problem and we can use the equation

$$V_1 M_1 = V_2 M_2$$

where V_1 = 500. mL = 0.500 L
             M_1 = 3.42 M = 3.42 mol/L
             M_2 = 0.0171 M = 0.0171 mol/L
             V_2 = Unknown final volume

Substituting into the equation and solving for V_2

$$V_2 = \frac{(0.500 \text{ L})(3.42 \text{ mol/L})}{(0.0171 \text{ mol/L})}$$
$$= 100. \text{ L}$$

13.   21.7 g Na_2CrO_4

The reaction equation is balanced as written and says that 1 mole of Pb^{2+} ion reacts with 1 mole of CrO_4^{2-} ion. So, we need to calculate how many moles of Pb^{2+} ion we have and from this calculate the mass of Na_2CrO_4 needed.

$$\text{moles of } Pb^{2+} = (\text{mass } Pb^{2+}/\text{molar mass } Pb^{2+}) \times 10.0 \text{ L}$$

$$= \left(\frac{2.78 \text{ g/L}}{207.2 \text{ g/mol}} \right)(10.0 \text{ L}) = 0.134 \text{ mol}$$

In order to precipitate 0.134 moles of Pb^{2+} ion, we must add an equal number of moles of Na_2CrO_4 which we can determine as follows:

molar mass of Na_2CrO_4 = 45.98 g + 52.00 g + 64.00 g + 162.0 g/mol
mass Na_2CrO_4 = (0.134 mol)(162.0 g/mol) = 21.7 g

14. (1) number   (6)   about twice
     (2) boiling point   (7)   2 moles
     (3) equal   (8)   about three
     (4) 1 kg   (9)   freezing point depression
     (5) vapor pressure lowering (10) boiling point elevation

15. (1) 6.67×10^2 g

The problem states that the temperature differential is from 0.0 °C to −20.0°C or a Δt of 20.0°. In addition, we have the value for K_f for water and the fact that we are dealing with 1 kilogram of solvent. We need to determine the molar mass of ethylene glycol and then use the equation given.

molar mass of $C_2H_4(OH)_2 = 2 \times 12.01$ g $= 24.02$ g
4×1.008 g $= 4.032$ g
2×16.01 g $= 32.00$ g
2×1.008 g $= 2.016$ g
$\overline{ 62.07 \text{ g/mol}}$

$$\text{g solute} = \frac{\Delta t_f \times \text{molar mass} \times \text{kg solvent}}{K_f}$$

Making the appropriate substitutions:

$$= \frac{(20.0°\cancel{C})(62.07 \text{ g/}\cancel{mol})(1 \text{ kg solvent})}{1.86°\cancel{C}/\cancel{mol}/\cancel{kg}}$$

$$= 6.67 \times 10^2 \text{ g of ethylene glycol}$$

(2) 140 g/mol

We are asked in this problem to determine the molar mass of a compound from freezing point depression data. In order to solve the equation, we need to express the amount of solvent in units of kilograms and then make the necessary substitutions of the other values.

20.0 g of benzene = 0.0200 kg
$K_f = 5.1°$C/mol/kg
$\Delta t_f = 2.75°$
grams of solute = 1.50 g

$$\text{molar mass} = \frac{K_f \times \text{g solute}}{\Delta t_f \times \text{kg solvent}}$$

$$= \frac{(5.1°\cancel{C})(1.50 \text{ g})}{\left(\dfrac{\text{mol}}{\text{kg}}\right)(2.75°\cancel{C})(0.0200 \text{ kg})}$$

$$= 140 \text{ g/mol}$$

16. Balanced equation is $3 \text{ Cd}^{2+}(aq) + 2 \text{ K}_3PO_4(aq) \rightarrow Cd_3(PO_4)_2(s) + 6 \text{ K}^+(aq)$. Volume of 5.00 M of K_3PO_4 required is 4.86 mL. The equation states that 3 moles of Cd^{2+} require 2 moles of K_3PO_4 to precipitate. The number of moles of Cd^{2+} ion present is equal to

$$\frac{4.10 \text{ g Cd}^{2+}}{112.4 \text{ g/mol}} = 0.0365 \text{ mol Cd}^{2+}$$

We will need the following number of moles of K_3PO_4.

$$\text{Moles K}_3PO_4 = (0.0365 \cancel{\text{ mol Cd}^{2+}}) \frac{2 \text{ mol K}_3PO_4}{3 \cancel{\text{ mol Cd}^{2+}}} = 0.0243 \text{ mol}$$

We next use dilution equation $V_1M_1 + V_2M_2$ where $V_1 =$ unknown volume of stock 5.00 M solution and

$M_1 = 5.00$ M.

Since 250. mL of the diluted K_3PO_4 is added to 600. mL of the Cd^{2+} solution, the final volume (V_2) will equal 850. mL (0.850 L).

$$V_2 = 250. \text{ mL} + 600. \text{ mL} = 0.850 \text{ L}$$
$$M_2 = \frac{0.0243 \text{ mol}}{0.850 \text{ L}} = 0.0286 \text{ M}$$
$$V_1 = \frac{(0.850 \text{ L})(0.0286 \text{ M})}{5.00 \text{ M}}$$
$$= 0.00486 \text{ L} = 4.86 \text{ mL}$$

17. Mass percent of glucose solution is 47.6%; molality is 5.05 m. Blood glucose molality is 0.50 m. The molar mass of glucose is 180.2 g/mol. A solution of 1.00 g glucose is 1.10 mL of H_2O is equal to a mass percent concentration of

$$\frac{1.00 \text{ g glucose}}{(1.10 \text{ g}) + (1.00 \text{ g})} \times 100 = 47.6\,\%$$

We can dissolve $\dfrac{1000.}{1.10}$ g of glucose in 1 kg of water.

This amount is equal to 909 g of glucose or $\dfrac{909 \text{ g}}{180.2 \text{ g/mol}} = 5.04 \text{ mol}$

The molality is therefore 5.04 m.

Human blood has a concentration of 0.090% or 0.90 g per 1000 g of blood. Using the assumption that 1000 g of blood is equivalent to 1000 g of solvent, the molality can be calculated as follows:

$$\frac{0.90 \text{ g}}{180.2 \text{ g/mol}} = 0.0050 \text{ mol}$$
$$\text{molality} = \frac{0.0050 \text{ mol}}{1 \text{ kg solvent}} = 0.0050 \text{ } m$$

CROSSWORD PUZZLE 1

Across

2. A mixture without a uniform composition.
6. Yellow solid nonmetallic element (atomic number 16).
7. An element that is ductile and malleable is a _____.
10. One shouldn't do crossword puzzles with a _____.
11. Binary compounds with a metallic element have names ending in _____.
12. Number of protons in hydrogen nucleus.
13. Chemical symbol for tantalum.
15. Chemical symbol for cobalt.
16. Elements whose *d* and *f* orbitals are being filled.
17. Formula for nitrogen oxide.
18. Light energy is sometimes considered to exist as a _____.
20. Chemical symbol for element number 95.
22. A pair of electrons forms a chemical _____ between two atoms.
24. Chemical symbol for osmium.
26. A positively charged ion.
27. Chemical symbol for element number 10.

Down

1. The answer to a mathematical problem is often called the _____.
3. Oxygen is an example of an _____, the basic building blocks of matter.
4. Nitrogen exists in what physical state as the uncombined element.
5. Chemical symbol for first element in period number 3.
8. Heat and light are forms of _____.
9. Element number 54.
10. Chemical symbol for element number 84.
13. The element whose chemical symbol is Sn.
14. The smallest indivisible unit of matter.
15. When elements combine chemically, they form a new substance called a _____.
19. A negatively charged ion is called an _____.
21. One Avogadro's Number of a substance dissolved in enough water to give 1 liter of solution is a 1 _____ solution.
23. Chemical symbol for element number 28.
25. Chemical symbol for element number 34.

Answers are found at the end of Chapter 20.

CHAPTER FIFTEEN

Acids, Bases, and Salts

SELECTED CONCEPTS REVISITED

Acids and bases have been defined in three different ways (and by three different sets of scientists whose names are associated with the different definitions). Note that these definitions simply supplement each other. In this course, you will most likely be dealing mainly with the Bronsted-Lowry definition.

Arrhenius acid – a substance that releases H^+ into solution.
Bronsted-Lowry acid – a proton donor
Lewis acid – an electron-pair acceptor
Arrhenius base – a substance that releases OH^- into solution
Bronsted-Lowry base – a proton acceptor
Lewis base – an electron-pair acceptor

An acid and its conjugate base are different structurally only by an H^+ (e.g. HCl and Cl^-). The same is true for a base and its conjugate acid (e.g. OH^- and H_2O).

Electrolytes are substances capable of producing ions when dissolved in water either by dissociation or ionization. The terms strong and weak describe qualitatively how many ions are produced per molecule or unit of the substance. Acids and bases are also described using the strong and weak terminology. For strong electrolytes, the generic equation $MX \rightarrow M^+ + X^-$ essentially goes to completion. For weak electrolytes, the reaction is written as an equilibrium reaction and the equilibrium lies to the left, for example, $MX \rightleftharpoons M^+ + X^-$.

The higher the H^+ concentration the more acidic the solution and therefore the lower the pH. At a pH of 7, the $[H^+] = [OH^-]$. A pH of less than 7 means that $[H^+] > [OH^-]$ and the solution is acidic.

When writing ionic equations the idea is to show in what form the majority of each species actually exists in solution. For example, since acetic acid (CH_3CO_2H) is a weak acid, when a sample of acetic acid is dissolved in water, most of it is in the molecular form with only a very small percentage in the ionic form. Therefore, acetic acid appears as CH_3CO_2H in a net ionic equation. However, when HCl (g) is dissolved in water, all the HCl ionizes to form H^+ and Cl^- and therefore HCl (aq) would be represented by H^+ and Cl^- in an ionic equation.

COMMON PITFALLS TO AVOID

Salt does not necessarily mean "table salt" or NaCl. Many different ionic compounds may be referred to as salts, that is, they can be made from the reaction of an acid with a base.

Pure water is not considered to be an electrolyte and is not considered to be ionic. Puddles, lakes and pools tend to be very good conductors of electricity but that is because of the impurities (of which some are ionic) dissolved in the water.

Do not forget to determine the number of ions in solution per unit or per molecule when dealing with the colligative properties of ionic solutions. Colligative properties depend on the number of particles in solution and each ion counts as a particle.

Not all titrations result in a solution of pH 7. The pH of the solution once the titration is complete depends on the acid and the base used. If a strong acid and a strong base are reacting then the end pH is usually 7, however if a weak acid or base is involved, the end pH is generally not 7.

Please make sure you are familiar with how to use the log and inverse log functions on your calculator.

Charges and atoms must be balanced for all properly balanced chemical equations.

Remember that each ionic compound (the ones you will meet) is generally composed of only one type of cation and one type of anion. If you find yourself breaking down an ionic compound into more than two different ions, you have made a mistake. If there are more than two elements present in the compound, chances are high that a poly-atomic ion is involved.

Use your state symbols to help determine which compounds get broken down into ions for ionic equations. Only compounds followed by (aq) have the potential to be broken down. Molecules followed by (s), (l), or (g) remain intact and are not written as ions.

When dealing with significant figures for logs, remember that the number in front of the decimal in the pH does not count as a significant figure.

SELF-EVALUATION SECTION

1. There are several different definitions of acids and bases according to various theories. Match the chemist's name with the appropriate phrase:

    (1) Arrhenius   a.   Acid is proton donor

    (2) Brønsted-Lowry   b.   Acid is electron pair acceptor

    (3) Lewis   c.   Acids are hydrogen-containing substances that dissociate to produce H^+

2. Identify the conjugate acid-base pairs in the following equations.

    (1) $H_2BO_3^- + H_3O^+ \rightarrow H_3BO_3 + H_2O$

    (2) $NH_3 + HBr \rightarrow NH_4^+ + Br^-$

    (3) $HSO_4^- + H_3O^+ \rightarrow H_2SO_4 + H_2O$

    (4) $HC_2H_3O_2 + H_2O \rightarrow H_3O^+ + C_2H_3O_2^-$

    (5) $NH_4^+ + H_2O \rightarrow NH_3 + H_3O^+$

3.   Complete and balance the following reactions between HCl and the chemicals given:

   (1)   HCl and NH_4OH

   (2)   HCl and Zn metal

   (3)   HCl and CaO

   (4)   HCl and Na_2CO_3

4.   Complete the equation for the reaction of H_3PO_4 and $Al(OH)_3$, an amphoteric hydroxide.

   $$Al(OH)_3 + H_3PO_4 \rightarrow$$

5.   Neutralization reactions. Identify the reactants as either *acids* or *bases*.

   (1) $NH_4OH + HClO_3 \rightarrow NH_4ClO_3 + H_2O$

   _____   _____

   (2) $3\ HCl + Al(OH)_3 \rightarrow AlCl_3 + 3\ H_2O$

   _____   _____

   (3) $H_2SO_4 + Ca(OH)_2 \rightarrow CaSO_4 + 2\ H_2O$

   _____   _____

   Dissociation reactions. Identify the acids and bases in each reaction according to the Brønsted-Lowry theory. Examine both the reactants and the products.

   (4) $NH_4^+ + H_2O \rightarrow NH_3 + H_3O^+$

   _____ _____ _____ _____

   (5) $C_2H_3O_2^- + H_2O \rightarrow HC_2H_3O_2 + OH^-$

   _____ _____ _____ _____

   (6) $Al^{3+} + H_2O \rightarrow Al(OH)^{2+} + H^+$

   _____ _____ _____ _____

6.   In which of the following situations does the solution contain an electrolyte (E) or nonelectrolyte (NE)?

   (1)   The solution contains an acid, such as H_2SO_4.   _____

   (2)   The solution is a nonconductor of electric current.   _____

   (3)   The solution contains only electrically neutral molecules.   _____

   (4)   The solution contains molecules, such as acetic acid, which have reacted with water to produce ions.   _____

   (5)   The solution contains gaseous oxygen.   _____

   (6)   The solution conducts an electric current.   _____

   (7)   The solution contains electrically charged ions.   _____

7. Identify the following compounds as strong electrolytes (s) or weak electrolytes (w).

   (1) HF   (6) HCl

   (2) $MgNO_3$ (7) HClO

   (3) NaOH   (8) H_3BO_3

   (4) $HC_2H_3O_2$ (9) $HClO_4$

   (5) $CaCl_2$ (10) NH_4OH

8. Calculate the molarity of the ions in each of the salt solutions listed below. Consider each salt to be 100% dissociated.

   (1) 0.25 M $MgCl_2$ (4) 0.035 M $Ca(NO_3)_2$
   (2) 0.80 M NH_4NO_3 (5) 1.2 M $Al_2(SO_4)_3$
   (3) 2.1 M K_3PO_4

9. Identify the following reactions as either dissociation (D) or ionization (I).

   (1) $NaCl \xrightarrow{H_2O} Na^+ + Cl^-$ _____

   (2) $HC_2H_3O_2 + H_2O \rightarrow H_3O^+ + C_2H_3O_2^-$ _____

   (3) $NH_3 + H_2O \rightarrow NH_4^+ + OH^-$ _____

   (4) $Ca(OH)_2 \xrightarrow{H_2O} Ca^{2+} + 2\ OH^-$ _____

   (5) $HCl + H_2O \rightarrow H_3O^+ + Cl^-$ _____

10. Rank these common solutions from strong acid to strong base. Indicate which solutions are acid, close to neutral, and basic using Figure 15.3.

	Solution	pH	Rank in order from acid to base
(1)	0.1 M HCl	1.0	
(2)	Blood	7.4	
(3)	Lemon juice	2.3	
(4)	Drain cleaner	12.1	
(5)	Vinegar	2.8	
(6)	0.1 M $HC_2H_3O_2$	2.9	
(7)	Milk	6.6	
(8)	Ammonium cleaner	8.2	
(9)	Carbonated water	3.0	
(10)	Tomato juice	4.1	

11. Let's do a little bit more with pH. The mathematical expression for pH indicates that we use the hydrogen ion concentration. For example, when $[H^+] = 1 \times 10^{-3}$ mol/L, the pH is 3.

    When the $[H^+] = 1 \times 10^{-9}$ mol/L, the pH is 9.

As you can see, we derive the pH value from the exponent of 10 when the number in front is 1. When the number in front of the exponent is between 1 and 10, the pH is between the power of 10 given and the next lower power.

As an example

$[H^+] = 6 \times 10^{-7}$ mol/L   pH is between 7 and 6
$[H^+] = 5.4 \times 10^{-2}$ mol/L pH is between 2 and 1

(1)   What is the pH of a solution whose $[H^+] = 1 \times 10^{-10}$ mol/L?

pH = _____

(2)   What is the pH of a solution whose $[H^+] = 1 \times 10^{-7}$ mol/L?

pH = _____

(3)   What is the pH of a solution whose $[H^+] = 4 \times 10^{-8}$ mol/L?

pH = _____

(4)   What is the pH of a solution whose $[H^+] = 3.3 \times 10^{-5}$ mol/L?

pH = _____

12.   Indicate whether the following equations are *formula* equations (F), *total ionic* equations (T) or *net ionic* equations (N).

(1) $H_2SO_{4(aq)} + MgCO_{3(aq)} \rightarrow MgSO_{4(aq)} + H_2O_{(l)} + CO_{2(g)}$ _____

(2) $2\,H^+_{(aq)} + 2\,Cl^-_{(aq)} + Na_2O_{(s)} \rightarrow 2\,Na^+_{(aq)} + 2\,Cl^-_{(aq)} + H_2O_{(l)}$ _____

(3) $H^+ + OH^- \rightarrow H_2O$ _____

(4) $2\,HCl_{(aq)} + Ca_{(s)} \rightarrow H_{2(g)} + CaCl_{2(aq)}$ _____

(5) $2\,H^+_{(aq)} + SO_4^{2-}{}_{(aq)} + Mg_{(s)} \rightarrow H_{2(g)} + Mg^{2+}_{(aq)} + SO_4^{2-}{}_{(aq)}$ _____

(6) $2\,OH^-_{(aq)} + Mn^{2+}_{(aq)} \rightarrow Mn(OH)_{2(s)}$ _____

(7) $2\,H^+_{(aq)} + CO_3^{2-}{}_{(aq)} \rightarrow H_2O_{(l)} + CO_{2(g)}$ _____

(8) $2\,OH^-_{(aq)} + Zn_{(s)} \rightarrow ZnO_2^{2-}{}_{(aq)} + H_{2(g)}$ _____

(9) $Zn(OH)_{2(s)} + 2\,H^+_{(aq)} + 2\,Cl^-_{(aq)} \rightarrow Zn^{2+}_{(aq)} + 2\,Cl^-_{(aq)} + 2\,H_2O_{(l)}$ _____

(10) $3\,NH_4OH_{(aq)} + FeCl_{3(aq)} \rightarrow Fe(OH)_{3(s)} + 3\,NH_4Cl_{(aq)}$ _____

13. Write balanced *formula, total ionic,* and *net ionic* equations for the following two equations written in words.

   (1) Barium chloride solution plus silver nitrate solution react to form insoluble silver chloride plus barium nitrate solution.

   Formula equation

   Total ionic equation

   Net ionic equation

   (2) Solid iron plus copper(II) sulfate solution react to form solid copper plus iron(II) sulfate solution.

   Formula equation

   Total ionic equation

   Net ionic equation

14. 35 mL of 0.20 M HCl is required to titrate 50. mL of $Ca(OH)_2$ solution. What is the molarity of the base?

   Do your calculations here.

Challenge Problems

15. Identify the conjugate acid-base pairs in the three step ionization of phosphoric acid, H_3PO_4. Write each equation and indicate the acid-base pairs.

    Do your calculations here.

16. How many grams, theoretically, of $BaSO_4$ will be precipitated when 35 mL of 0.15 M H_2SO_4 is added to 18 mL of 0.43 M $Ba(OH)_2$? What is the limiting reactant? If the actual yield is 1.0 g $BaSO_4$, what is the percent yield?

    Do your calculation here.

17. You are an analyst working on product quality for a vinegar company. Your work this morning involves 10.00 mL samples of various batches of wine vinegar. The data table for the four titrations looks like this:

    Sample volume: 10.00 mL

	Batch A	B	C	D
Volume of 0.2500 M NaOH	33.05 mL	34.75 mL	32.80 mL	34.35 mL

    If the acidity is due to acetic acid $(HC_2H_3O_2)$ what is the average molarity of the vinegar? Given that the density of the vinegar is 1.005 g/mL, what is the average mass percent concentration?

RECAP SECTION

Chapter 15 is an excellent chapter to be related to your laboratory experiments. You will be working with acids and bases, doing titrations, using the pH system, and examining the properties of electrolytes, nonelectrolytes, and colloids. Each of these topics is an important part of Chapter 15, and laboratory experiments will help clarify what was covered in lectures and homework problems. The last section on equation writing will be carried over to Chapter 17 when you examine oxidation-reduction reactions and try your hand at balancing all types of reaction equations.

1. (1)   c   (2) a   (3) b

2. (1) $H_2BO_3^-$ base, H_3BO_3 acid; H_3O^+ acid, H_2O base
   (2) NH_3 base, NH_4^+ acid; HBr acid, Br^- base
   (3) HSO_4^- base, H_2SO_4 acid; H_3O^+ acid, H_2O base
   (4) $HC_2H_3O_2$ acid, $C_2H_3O_2^-$ base; H_2O base, H_3O^+ acid
   (5) NH_4^+ acid, NH_3 base; H_2O base, H_3O^+ acid

3. (1) $HCl_{(aq)} + NH_4OH_{(aq)} \rightarrow NH_4Cl_{(aq)} + H_2O_{(l)}$
   (2) $2\,HCl_{(aq)} + Zn_{(s)} \rightarrow ZnCl_{2(aq)} + H_{2(g)}$
   (3) $2\,HCl_{(aq)} + CaO_{(s)} \rightarrow CaCl_{2(aq)} + H_2O_{(l)}$
   (4) $2\,HCl_{(aq)} + Na_2CO_{3(aq)} \rightarrow 2\,NaCl_{(aq)} + H_2O + CO_{2(g)}$

4. $Al(OH)_{3(s)} + H_3PO_{4(aq)} \rightarrow AlPO_{4(s)} + 3\,H_2O_{(l)}$

5. (1) NH_4OH = base; $HClO_3$ = acid
   (2) $Al(OH)_3$ = base;   HCl = acid
   (3) H_2SO_4 = acid; $Ca(OH)_2$ = base
   (4) NH_4^+ = acid; H_2O = base; NH_3 = base; H_3O^+ = acid
   (5) $C_2H_3O_2^-$ = base; H_2O = acid; $HC_2H_3O_2$ = acid; OH^- = base
   (6) Al^{3+} = acid; H_2O = base; $Al(OH)^{2+}$ = base; H^+ = acid

6. (1)   E   (2)   NE   (3)   NE   (4)   E   (5)   NE   (6)   E   (7)   E

7. (1)   w   (2)   s   (3)   s   (4)   w   (5)   s
   (6)   s   (7)   w   (8)   w   (9)   s   (10)   w

8. (1)   0.25 M Mg^{2+}, 0.50 M Cl^- (4)   0.035 M Ca^{2+}, 0.070 M NO_3^-
   (2)   0.80 M NH_4^+, 0.80 M NO_3^- (5)   2.4 M Al^{3+}, 3.6 M SO_4^{2-}
   (3)   6.3 M K^+, 2.1 M PO_4^{3-}

9. (1)   Dissociation (D). Ions are present in the crystalline salt, and the water molecules act only as the sepa-
         ration agent. Molten salt as well as a solution of salt conducts electricity.

   (2)   Ionization (I). Water molecules are necessary to form ions in reaction with $HC_2H_3O_2$. Pure $HC_2H_3O_2$ is
         a very weak electrolyte.

   (3)   Ionization (I). When ammonia gas dissolves in water, it undergoes reaction with water molecules to
         form a solution we call ammonium hydroxide (NH_4OH).

   (4)   Dissociation (D).

   (5)   Ionization (I). The bond in $HCl_{(g)}$ is predominately covalent.

10. (1)   (3)   (5)   (6)   (9)   (10)   (7)   (2)   (8)   (4)
    The first six (in order above) are acids, next two close to neutral, and the last two are basic.

11. (1)   10   (2)   7   (3)   Between 8 and 7   (4)   Between 5 and 4

— 152 —

12. (1) F (2) T (3) F (4) F (5) T
    (6) N (7) N (8) N (9) T (10) F

13. (1) Formula (F) $BaCl_{(aq)} + 2\ AgNO_{3(aq)} \rightarrow 2\ AgCl_{(s)} + Ba(NO_3)_{2(aq)}$

    Total ionic (T) $Ba^{2+}_{(aq)} + 2\ Cl^-_{(aq)} + 2Ag^+_{(aq)} + 2\ NO_3^-_{(aq)} \rightarrow$
    $$2\ AgCl_{(s)} + Ba^{2+}_{(aq)} + 2NO_3^-_{(aq)}$$

    Net ionic (N) $Ag^+_{(aq)} + Cl^-_{(aq)} \rightarrow AgCl_{(s)}$

    (2) Formula (F) $Fe_{(s)} + CuSO_{4(aq)} \rightarrow Cu_{(s)} + FeSO_{4(aq)}$

    Total Ionic (T) $Fe_{(s)} + Cu^{2+}_{(aq)} + SO_{4(aq)}^{2-} \rightarrow Cu_{(s)} + Fe^{2+}_{(aq)} + SO_{4(aq)}^{2-}$

    Net ionic (N) $Fe_{(s)} + Cu^2_{(aq)} \rightarrow Cu_{(s)} + Fe^{2+}_{(aq)}$

14. 0.070 M. Equation is $2\ HCl_{(aq)} + Ca(OH)_{2(aq)} \rightarrow CaCl_{2(aq)} + 2\ H_2O_{(l)}$. The number of moles of acid used is twice the number of moles of base. Therefore, if we calculate the number of moles of acid used we can use the mole-ratio method to determine the number of moles of $Ca(OH)_2$ in the 50. mL volume, and thus the concentration or molarity.

$$\text{number of moles of HCl} = 0.035\ \cancel{L} \times \frac{0.20\ \text{mol}}{\cancel{L}}\ HCl$$
$$= 0.0070\ \text{mol}$$

the number of moles of $Ca(OH)_2$ is
$$0.0070\ \cancel{\text{mol HCl}} \times \frac{1\ \text{mol } Ca(OH)_2}{2\ \cancel{\text{mol HCl}}} = 0.0035\ \text{mol } Ca(OH)_2$$

Therefore, 0.0035 mol of $Ca(OH)_2$ was present in 50. mL of $Ca(OH)_2$ solution.

Molarity of $Ca(OH)_2$
$$M = \frac{\text{mol}}{L} = \frac{0.0035\ \text{mol } Ca(OH)_2}{0.050\ L} = 0.070\ M\ Ca(OH)_2$$

15. (1) $H_3PO_4 + H_2O \rightarrow H_3O^+ + H_2PO_4^-$

    (2) $H_2PO_4^- + H_2O \rightarrow H_3O^+ + HPO_4^{2-}$

    (3) $HPO_4^{2-} + H_2O \rightarrow H_3O^+ + PO_4^{3-}$

In each case H_3O^+ and H_2O are a conjugate acid-base pair. For reaction (1) H_3PO_4 is the acid and $H^2PO_4^-$ is the base. For reaction (2) $H_2PO_4^-$ is the acid and HPO_4^{2-} is the base while in (3) HPO_4^{2-} is the acid and PO_4^{3-} is the base.

16. H_2SO_4 is limiting, theoretical yield is 1.2 g $BaSO_4$, percent yield is 83%.

    Net ionic equation is $Ba^{2+} + SO_4^{2-} \rightarrow BaSO_{4(s)}$

    Number of moles $Ba^{2+} = \left(0.43\ \frac{\text{mol}}{\cancel{L}}\right)(0.018\ \cancel{L}) = 0.0077\ \text{mol}$

    Number of moles $SO_4^{2-} = \left(0.15\ \frac{\text{mol}}{\cancel{L}}\right)(0.035\ \cancel{L}) = 0.0053\ \text{mol}$

    Molar mass $BaSO_4 = 233\ \frac{g}{\text{mol}}$

$$\text{Amount BaSO}_4 = \left(233 \, \frac{g}{mol}\right)(0.0053 \, mol) = 1.2 \, g$$

$$\text{Percent yield} = \frac{\text{Actual yield}}{\text{Theoretical yield}} \times 100 = \frac{1.0 \, g}{1.2 g} \times 100 = 83\%$$

17. $NaOH + HC_2H_3O_2 \rightarrow H_2O + NaC_2H_3O_2$

From the volume of NaOH used and the molarity, the moles of NaOH used may be determined for each batch.

molarity $\times$ volume (in liters) = moles

	Batch A	B	C	D
moles NaOH used	0.008263	0.008688	0.008200	0.008588

Since the mole ratio NaOH to $HC_2H_3O_2$ is 1:1, the moles of acetic acid in each batch is the same as the moles of NaOH used. Therefore the average number of moles of acetic acid can be calculated.

average moles $HC_2H_3O_2$ = 0.008435 moles

Using the sample volume of 10.00 ml, the average molarity of the vinegar can be calculated.

$$\text{molarity} = \frac{mol}{L} = 8\frac{0.008435 \, mol}{0.01000L}$$

Molarity of $HC_2H_3O_2$ = 0.8435 M

The number of grams of acetic acid can be calculated from the molarity relationship.

$$0.8435 \, \frac{mol}{L} \times 60.06 \, \frac{g}{mol} = 50.66 \, g/L$$

If the density of the vinegar is 1.005 g/mL then 1000 mL of vinegar will have a mass of:

1.005 g/mL $\times$ 1000 mL = 1005 g

The mass percent calculation is

$$\frac{50.66 \, g}{1005 \, g} \times 100 = 5.041\% \, HC_2H_3O_2$$

CHAPTER SIXTEEN

Chemical Equilibrium

SELECTED CONCEPTS REVISITED

A reversible reaction is a reaction that can proceed in either direction. That is, as written the reactants may proceed to products or the products can react to reform the reactants.

A reaction has reached equilibrium when the rate of the forward reaction is equal to the rate of the reverse reaction.

Le Chatelier's Principle basically states that a reaction at equilibrium will shift its equilibrium position to relieve any added stress. For example, if more reactant is added to the reaction at equilibrium, the reaction will proceed forwards so as to use up some of the added reactant and so the position of the equilibrium will be shifted to the right.

$$K = \frac{[products]}{[reactants]}$$

Equilibrium constants (K) indicate numerically the position of the equilibrium by providing a ratio of the [products] to the [reactants]. The larger the constant, K, the larger the amount of products present relative to reactants and therefore equilibrium lies further to the right.

The equilibrium constants of certain types of equilibrium reactions have been given special names. Some of these are listed below. You should realize that they are still all simply a ratio of the products to the reactants.

K_w = ion product constant for water (the equilibrium constant for the autoionization of water).
K_a = acid dissociation constant (the equilibrium constant for the dissociation or ionization of an acid in water).
K_{sp} = the equilibrium constant for a slightly soluble salt.

The effect caused by the addition of an ion already in a solution at equilibrium is called the common ion effect (the equilibrium position is shifted in the direction that uses the ion as a reactant). Due to the common ion effect, a buffered solution resists changes in pH when small amounts of either acid or base are added to the solution.

COMMON PITFALLS TO AVOID

A reaction at equilibrium has not stopped. Both the forward and the reverse reaction are actually taking place but at the same rate so that no overall change over time is noticed.

A reaction at equilibrium does not necessarily have equal amounts of reactants to products. In fact the equilibrium position for most reactions is not in the middle. Equilibrium means the rate of both forwards and reverse reactions are equal. The equilibrium position gives information as to the relative concentrations of reactants to products.

The addition or removal of a catalyst does NOT affect the equilibrium position. A catalyst simply affects how fast the reaction reaches equilibrium but does not affect the equilibrium constant.

SELF-EVALUATION SECTION

Most of the examples and problems that we examine for Chapter 16 will have close counterparts in the text material. You should refer to the examples in the text as needed.

1. Write the following word equations for reversible systems in chemical equation form.

    (1) Four moles of hydrogen chloride gas plus one mole of oxygen gas react to produce two moles of steam (gaseous water) and two moles of chlorine gas.

    _____

    (2) One mole of chlorine gas reacts with one mole of liquid water to produce one mole of HClO solution and one mole of hydrochloric acid solution.

    _____

    (3) One mole of carbon in the form of coke plus one mole of carbon dioxide gas plus heat react to produce two moles of carbon monoxide gas.

    _____

    (4) One mole of nitrogen gas plus three moles of hydrogen gas react to produce two moles of ammonia gas plus heat.

    _____

    (5) Two moles of sulfur dioxide gas plus one mole of oxygen gas react to produce two moles of sulfur tri-oxide gas.

    _____

2. Using the Principle of Le Chatelier, predict the effect of changing various conditions on the following equilibrium systems.

    (1) $2 SO_{2(g)} + O_{2(g)} \rightleftarrows 2 SO_{3(g)}$

    Which direction, right or left, will the equilibrium be shifted if:

    a. the amount of SO_2 is decreased? _____

    b. the pressure is increased? _____

    c. a catalyst is added? _____

(2) $3 O_{2(g)} + heat \rightleftharpoons 2 O_{3(g)}$

Which direction will the equilibrium be shifted if:

a. the amount of ozone is decreased? _____

b. the amount of oxygen is increased? _____

c. the reaction is cooled? _____

d. the pressure is decreased? _____

(3) $C_{(s)} + CO_{2(g)} \rightleftharpoons 2 CO_{(g)}$

Which direction will the equilibrium be shifted if:

a. the amount of carbon is increased? _____

b. the reaction mixture is heated? _____

c. the amount of CO is increased? _____

d. the amount of CO_2 is increased? _____

e. the pressure is increased? _____

3. Write the equilibrium constant expression for the following reactions.

(1) $PbSO_{4(s)} \rightleftharpoons Pb^{2+}_{(aq)} + SO_4^{2-}_{(aq)}$

(2) $Ag^+ + 2 NH_3 \rightleftharpoons Ag(NH_3)_2^+$

(3) $3 O_{2(g)} \overset{\Delta}{\rightleftharpoons} 2 O_{3(g)}$

(4) $C_{(s)} + CO_{2(g)} \overset{\Delta}{\rightleftharpoons} 2 CO_{(g)}$

(5) $CH_{4(g)} + 2 O_{2(g)} \rightleftharpoons CO_{2(g)} + 2 H_2O_{(g)}$

(6) $HNO_2 \rightleftharpoons H^+ + NO_2^-$

(7) $Al^{3+} + OH^- \rightleftharpoons Al(OH)^{2+}$

(8) $2 SO_{2(g)} + O_{2(g)} \rightleftharpoons 2 SO_{3(g)}$

(9) $4 NH_{3(g)} + 5 O_{2(g)} \rightleftharpoons 4 NO_{(g)} + 6 H_2O_{(g)}$

Write your answers here.

4. What is the value of K_{eq} for the reaction shown if a 1.0 L flask originally contains 0.5 moles of $SO_{3(g)}$ at 832 °C and, at equilibrium, 50 percent of the SO_3 gas has decomposed?

$$2 SO_{3(g)} \rightleftharpoons 2 SO_{2(g)} + O_{2(g)}$$

5. (1) What is the concentration of hydrogen ions, H^+, in a 0.25 M solution of nitrous acid, HNO_2? $K_a = 4.5 \times 10^{-4}$

Do your calculations here.

(2)   What is the concentration of hydrogen ions, H^+, in a 0.15 M solution of butyric acid, $CH_3(CH_2)_2COOH$? $K_a = 1.52 \times 10^{-5}$

Do your calculations here.

6.   (1)   What is the H^+ ion concentration and the OH^- ion concentration in a solution of pH 6?

Do your calculations here.

(2)   What is the pH of a 1×10^{-5} M KOH solution? Assume that the KOH is 100% ionized.

Do your calculations here.

(3)   What is the H^+ ion concentration and the OH^- ion concentration in a 0.01 M HBr solution? Assume the HBr is 100% ionized. $K_w = 1 \times 10^{-14}$

Do your calculations here.

(4)   Let's review in a short drill the relationships between pH, pOH, H^+, and OH^- concentrations.

a.   The pH of a solution is 8.

What is the pOH? _____

What is the H^+ concentration? _____

What is the OH^- concentration? _____

Is the solution acidic or basic? _____

b.   The pOH of a solution is 10.

What is the OH^- concentration? _____

What is the pH? _____

What is the H^+ concentration? _____

Is the solution acidic or basic? _____

c.   The H^+ concentration is 1×10^{-3} mol/L.

What is the pH? _____

What is the pOH? _____

What is the OH^- concentration? _____

Is the solution acidic or basic? _____

7.   (1)   Which of the following acids would be better used to make a buffer solution? H_2SO_4 or $HC_3H_5O_2$

(2)   Why?

(3)   What would be an ideal second component of the buffer?

(4)   Write the equilibrium equation between the two components of the buffer solution.

(5)   With which ion (or molecule) in the equation just written would added H^+ react with? Would added OH^- react with?

8.   Look at the compounds listed below. Match up the pairs that would be used for various buffer solutions.

a. $HC_2H_3O_2$ d. $H_2PO_4^-$

b. Na_2HPO_4 e. $NaC_2H_3O_2$

c. H_2CO_3 f. $NaHCO_3$

9.   (1)   Lead carbonate, $PbCO_3$, is a slightly soluble salt that dissociates according to the following reaction:

$$PbCO_{3(s)} \rightleftarrows Pb^{2+}{}_{(aq)} + CO_3{}^{2-}{}_{(aq)} \qquad K_{sp} = 3.3 \times 10^{-14}$$

What is the concentration of $CO_3{}^{2-}$ in a saturated lead carbonate solution?

Do your calculations here.

(2)   Strontium sulfate, $SrSO_4$, is a slightly soluble salt that dissociates according to the following reaction:

$$SrSO_{4(s)} \rightleftarrows Sr^{2+}_{(aq)} + SO_4^{2-}_{(aq)} \qquad K_{sp} = 3.8 \times 10^{-7}$$

What is the concentration of Sr^{2+} in a saturated $SrSO_4$ solution?

Do your calculations here.

10.   (1)   Calcium fluoride, CaF_2, has a solubility of 1.6×10^{-3} g per 100 mL of H_2O. Calculate the K_{sp} remembering that CaF_2 produces two F^- ions and one Ca^{2+} ion when it dissociates. The dissociation equation is

$$CaF_{2(s)} \rightleftarrows Ca^{2+}_{(aq)} + 2\,F^-_{(aq)}$$

Do your calculations here.

(2)   Tin (II) carbonate, $SnCO_3$, has a solubility of 5.73×10^{-3} g/L. Calculate the K_{sp}.

Do your calculations here.

(3)   One form of zinc sulfide, ZnS (sphalerite), has a solubility of 6.4×10^{-4} g/L. Calculate the K_{sp}.

Do your calculations here.

11.   Examine the formulas for the compounds shown. Make two lists ranking the acids from strongest to weakest and the salts from most soluble to least soluble. Then name all of the compounds.

HClO	$K_a = 3.5 \times 10^{-8}$
CaF_2	$K_{sp} = 3.9 \times 10^{-11}$
HCN	$K_a = 4.0 \times 10^{-10}$
$HC_2H_3O_2$	$K_a = 1.8 \times 10^{-5}$
$PbSO_4$	$K_{sp} = 1.3 \times 10^{-8}$
$BaSO_4$	$K_{sp} = 1.5 \times 10^{-9}$

12. Calculate the H^+ ion concentration and the pH in a 0.10 M H_3PO_4 solution assuming the first ionization reaction is the primary reaction.

$$H_3PO_4 \leftrightarrows H^+ + H_2PO_4^- \qquad K_a = 7.5 \times 10^{-3}$$

Do your calculations here.

Challenge Problems

13. At 720 K, the K_{eq} for the following reaction has a value of 50.5.

$$H_{2(g)} + I_{2(g)} \rightleftarrows 2 HI_{(g)}$$

At the given temperature, what is the value for K_{eq} for the reverse reaction.

$$2 HI_{(g)} \rightleftarrows H_{2(g)} + I_{2(g)}$$

Calculate the equilibrium concentrations of all 3 species given a 1 L container and 0.0300 moles of HI gas at 720 K.

Do your calculations here.

14. Calculate the pH of a buffer solution prepared by adding 0.10 mole of sodium acetate to 500 mL of a solution labeled 0.40 M acetic acid. K_a is 1.8×10^{-5}.

    Do your calculations here.

15. By using net ionic hydrolysis reactions, show why solutions of the following salts are either acidic or basic.

    (1) NH_4NO_3 – acidic

    (2) K_2CO_3 – basic

    (3) $NaCN$ – basic

    (4) $Al_2(SO_4)_3$ – acidic

16. Table 16.5 in the text shows that when 0.010 mol HCl or 0.010 mol NaOH is added to 1000 mL of water, the pH changes by 5 pH units. For each of the following 1000 mL mixtures would you expect a similar, large or small pH change when 0.010 mol NaOH is added.

    (1)   mixture made from equal amounts of 0.020 M $HClO_4$ and 0.020 M $NaClO_4$.

    (2)   mixture made from equal amounts of 0.020 M KNO_3 and 0.020 M HNO_3.

    (3)   mixture made from equal amounts of 0.040 M H_2SO_3 and 0.020 M NaOH.

    (4)   mixture made from equal amounts of 0.010 M HF and 0.010 M HBr.

RECAP SECTION

Chapter 16 discusses two of the most interesting topics in chemistry – chemical equilibrium and chemical kinetics. The material in the text is especially important for you to read and discuss with your instructor. The exercises in the study guide are primarily of a problem-solving nature; the concepts have been covered in the text. Once equilibrium has been reached in a chemical reaction system, you can use Le Chatelier's Principle to predict what will happen to the system if reaction conditions are altered. Such predictions have important practical value in the chemical industry. In addition, you have learned to use the equilibrium constant expression to evaluate chemical reactions for ion concentrations and equilibrium constants.

Scientists in industry and research draw on the concepts of Chapter 16 constantly to help them solve practical chemical problems.

1.   (1) $4\,HCl_{(g)} + O_{2(g)} \rightleftarrows 2\,H_2O_{(g)} + 2\,Cl_{2(g)}$

     (2) $Cl_{2(g)} + H_2O_{(l)} \rightleftarrows HClO_{(aq)} + HCl_{(aq)}$

     (3) $C_{(s)} + CO_{2(g)} + heat \rightleftarrows 2\,CO_{(g)}$

     (4) $N_{2(g)} + 3\,H_{2(g)} \rightleftarrows 2\,NH_{3(g)} + heat$

     (5) $2\,SO_{2(g)} + O_{2(g)} \rightleftarrows 2\,SO_{3(g)}$

2.   (1)   a. Decreasing the amount of SO_2 will produces a shift to the left. b. Increasing the pressure will cause a shift to the right, which is the side of fewer molecules. c. A catalyst will have no effect on the equilibrium.

     (2)   a. Decreasing the ozone concentration will shift the system to the right. b. So will increasing the amount of oxygen. c. Cooling the reaction will cause a shift to the left. d. Decreasing the pressure will cause a shift to the left.

     (3)   a. Increasing the amount of carbon will not shift the equilibrium. b. Heating the mixture will shift the equilibrium to the right. c. Increasing the amount of CO will shift the reaction to the left. d. Increasing the amount of CO_2 will produce the opposite effect. e. Increasing the pressure will drive the reaction to the left.

3.   (1) $K_{eq} = [Pb^{2+}][SO_4^{2-}]$

         $PbSO_4$ is a solid; therefore, it is left out of the K_{sp} expression.

     (2) $K_{eq} = \dfrac{[Ag(NH_3)_2^+]}{[Ag^+][NH_3]^2}$

     (3) $K_{eq} = \dfrac{[O_3]^2}{[O_2]^3}$

     (4) $K_{eq} = \dfrac{[CO]^2}{[CO_2]}$

         Carbon is in the solid state; therefore, it does not change concentration. It can be left out of the equilibrium expression.

     (5) $K_{eq} = \dfrac{[CO_2][H_2O]^2}{[CH_4][O_2]^2}$

     (6) $K_{eq} = \dfrac{[H^+][NO_2^-]}{[HNO_2]}$

     (7) $K_{eq} = \dfrac{[Al(OH)^{2+}]}{[Al^{3+}][OH^-]}$

(8) $\quad K_{eq} = \dfrac{[SO_3]^2}{[SO_2]^2[O_2]}$

(9) $\quad K_{eq} = \dfrac{[NO]^4[H_2O]^6}{[NH_3]^4[O_2]^5}$

4. $\quad K_{eq} = 0.13$

The expression for the equilibrium constant, K_{eq}, would be:

$$K_{eq} = \dfrac{[SO_2]^2[O_2]}{[SO_3]^2}$$

If one-half of the SO_3 has decomposed at equilibrium, we can write our equilibrium conditions in tabular form.

Substance	Initial Conditions	Final Conditions
SO_3	0.5 M	0.25 M
SO_2	0 M	0.25 M
O_2	0 M	$\dfrac{0.25}{2} M$

The equilibrium concentration of O_2 is one-half that of SO_2 according to the reaction. Substituting into the expression for K_{eq} we have

$$K_{eq} = \dfrac{[SO_2]^2[O_2]}{[SO_3]^2} = \dfrac{[0.25]^2[0.125]}{[0.25]^2} = 0.13$$

5. (1) $\quad 1.1 \times 10^{-2}$ M or mol/L

The formula for nitrous acid is HNO_2, so the first step is to write out the ionization equation. Then formulate the equilibrium expression.

$$HNO_2 \rightleftarrows H^+ + NO_2^-$$

$$K_a = \dfrac{[H^+][NO_2^-]}{[HNO_2]}$$

$$4.5 \times 10^{-4} = \dfrac{[H^+][NO_2^-]}{[HNO_2]}$$

Now you are ready to determine your strategy for making substitutions in the equation.

One way to keep everything straight is to set up a small table such as you did for gas law problems.

Substance	Initial Conditions	Final Conditions
H^+	0 M	X
NO_2^-	0 M	X
HNO_2	0.25 M	0.25 − X

Each molecule of HNO_2 that ionizes produces 1 H^+ ion and 1 NO_2^- ion. Therefore, let their equilibrium concentration be "X" so that the amount of HNO_2 left at equilibrium will be 0.25 M − X. Substituting in the equation

$$4.5 \times 10^{-4} = \frac{[X][X]}{[0.25 - X]}$$

If X is small compared with 0.25 M, it is possible to simply the equation.

$$4.5 \times 10^{-4} = \frac{X^2}{0.25}$$
$$1.1 \times 10^{-4} = X^2$$

To find the value for X, take the square root of 1.1 and 10^{-4}. Refer to a math text or ask your instructor if you have difficulty with square roots.

Therefore,

$$X = 1.1 \times 10^{-2} \, M \text{ or mol/L} = [H^+]$$

X is small compared to 0.25 so you were justified in making your assumption.

(2) 1.5×10^{-3} M or mol/L

The formula for butyric acid is $CH_3(CH_2)_2COOH$, so the first step is to write out the ionization equation. After that formulate the equilibrium expression. The ionizable hydrogen in this case is shown on the right.

$$CH_3(CH_2)_2COOH \rightleftarrows CH_3(CH_2)_2COO^- + H^+$$

$$K_a = \frac{[CH_3(CH_2)_2COO^-][H^+]}{CH_3(CH_2)_2COOH}$$

Set up a table as before.

Substance	Initial Conditions	Final Conditions
H^+	0 M	X
$CH_3(CH_2)_2COO^-$	0 M	X
$CH_3(CH_2)_2COOH$	0.15 M	0.15 − X

Substituting in our equilibrium expression, you have the following equation

$$1.52 \times 10^{-5} = \frac{[X][X]}{[0.15 - X]}$$

Assuming the X is small compared with 0.15, you can simplify your equation.

$$1.52 \times 10^{-5} = \frac{X^2}{0.15}$$
$$2.3 \times 10^{-6} = X^2$$

Taking the square root of both sides without difficulty is possible since the roots are divisible by 2. Therefore,

$$X = 1.5 \times 10^{-3} \, M \text{ or mol/L}$$

X is small compared to 0.15 so you were justified in making your assumption.

6. (1) $H^+ = 1 \times 10^{-6}$ M or mol/L, $OH^- = 1 \times 10^{-8}$ M or mol/L. We can determine the H^+ ion concentration directly from the pH and then use the ion product constant of water to calculate the OH^- ion concentration.

The pH value is a whole number, which means that the H^+ concentration is 1 times a negative amount of 10. The value of the exponent is the same as the pH value. Therefore, if the pH is 6, the H^+ ion concentration is 1×10^{-6} M or mol/L. Substituting this value into the ion product constant expression

$$[H^+][OH^-] = K_w = 1 \times 10^{-14}$$

$$[1 \times 10^{-6}][OH^-] = 1 \times 10^{-14}$$

$$[OH^-] = \frac{1 \times 10^{-14}}{1 \times 10^{-6}} = 1 \times 10^{-8} \text{ M or mol/L}$$

(2) pH = 9

There are at least two ways to solve this problem. By one method, we first need to find out what the H^+ concentration is. We can use the ion product constant expression to do this.

$$[H^+][OH^-] = K_w = 1 \times 10^{-14}$$

$$[H^+][1 \times 10^{-14}] = 1 \times 10^{-14}$$

$$[H^+] = \frac{1 \times 10^{-14}}{1 \times 10^{-5}} = 1 \times 10^{-9} \text{ M or mol/L}$$

We can easily convert this value to pH since the value in front of the exponential value is 1 (one). Therefore, the pH is 9. Another way to solve the problem is to use the relationship

$$pH + pOH = 14$$

Converting OH^- concentration into pOH, we have

$$1 \times 10^{-5} = \text{pOH of 5}$$

Then, we can substitute the pOH value

$$pH + 5 = 14$$
$$pH = 14 - 5 = 9$$

(3) $H^+ = 1 \times 10^{-2}$ M or mol/L, $OH^- = 1 \times 10^{-12}$ M or mol/L.
The problem states that the HBr is completely ionized. The equation would be

$$HBr \rightarrow H^+ + Br^-$$

This means that the H^+ concentration is the same as the HBr solution concentration or 0.01 M. In scientific notation, 0.01 M would be 1×10^{-2} M. To determine the OH^- concentration, we need to use the ion product constant of water relationship, which states that

$$[H^+][OH^-] = K_w = 1 \times 10^{-14}$$

We know what the concentration of H^+ ion is, so we can make the following substitution into the ion product constant equation

$$[1 \times 10^{-2}][OH^-] = 1 \times 10^{-14}$$

Therefore,

$$[OH^-] = \frac{1 \times 10^{-14}}{1 \times 10^{-2}} = 1 \times 10^{-12} \text{ M or mol/L}$$

Remember, when we are dividing exponents we subtract the denominator exponent from the numerator exponent.

$$-14 - (-2) = -14 + 2 = -12$$

(4)   a.   pOH = 6
$[H^+] = 1 \times 10^{-8}$ M
$[OH^-] = 1 \times {}^{-6}$ M
Basic

b. $[OH^-] = 1 \times 10^{-10}$ M
pH = 4
$[H^+] = 1 \times 10^{-4}$ M
Acidic

c.   pH = 3
pOH = 11
$[OH^-] = 1 \times 10^{-11}$ M
Acidic

7.   (1) $HC_3H_5O_2$

(2)   The best buffers are made from weak acids or weak bases. $HC_3H_5O_2$ is a weak acid; sulfuric acid is a strong acid.

(3)   A soluble salt of the weak acid would be ideal. For example, $NaC_3H_5O_2$.

(4) $HC_3H_5O_2 \rightleftarrows C_3H_5O_2^- + H_3O^+$

(5) $C_3H_5O_2^-$ would react with added H^+. $HC_3H_5O_2$ would react with any added OH^-.

8.   a and e, b and d, c and f.

9.   (1) $CO_3^{2-} = 1.8 \times 10^{-7}$ M or mol/L

First, write the equilibrium expression with the appropriate value for the solubility product

$$PbCO_3 \rightleftarrows Pb^{2+} + CO_3^{2-}$$
$$K_{sp} = 3.3 \times 10^{-14} = [Pb^{2+}][CO_3^{2-}]$$

Since $PbCO_3$ is a solid whose concentration is unchanging in the equilibrium system you can cancel the term $[PbCO_{3(s)}]$ out of the equation. Since the concentration of Pb^{2+} is equal to the concentration of CO_3^{2-} at equilbrium, you can substitute "X" into the K_{sp} equation for both ion concentrations.

$$3.3 \times 10^{-14} = [X][X]$$
$$3.3 \times 10^{-14} = X^2$$
$$1.8 \times 10^{-7} = X$$

Therefore,

$$[Pb^{2+}] = [CO_3^{2-}] = 1.8 \times 10^{-7} M$$

(2) $Sr^{2+} = 6.2 \times 10^{-4} M$
The equation is

$$K_{sp} = 3.8 \times 10^{-7} = [Sr^{2+}][SO_4^{2-}]$$

Since

$$[Sr^{2+}] = [SO_4^{2-}] = X$$

Therefore:

$$3.8 \times 10^{-7} = [X][X]$$
$$3.8 \times 10^{-7} = X^2$$
$$6.2 \times 10^{-4} = X$$

Therefore:

$$[Sr^{2+}] = [SO_4^{2-}] = 6.2 \times 10^{-4} M$$

10.   (1) $K_{sp} = 3.2 \times 10^{-11}$

First calculate the molar mass of CaF_2, using a list of atomic masses.

$$\text{molar mass of } CaF_2 = 40.08 \text{ g} + (2 \times 19.00 \text{ g}) = 78.08 \frac{g}{mol}$$

A solubility of 1.6×10^{-3} g per 100 mL of H_2O is the same as 1.6×10^{-2} g per L. Using this number, we can determine the molarity of a saturated solution of CaF_2, assuming the volume of water to be identical with the volume of the solution.

$$\text{Molarity} = \frac{\text{moles}}{L}$$

Substituting in the above equation.

$$M = \frac{1.6 \times 10^{-2} \text{ g/L}}{78.08 \text{ g/mol}} = 2.0 \times 10^{-4} M$$
$$CaF_{2(s)} \rightleftarrows Ca^{2+}_{(aq)} + 2 F^-_{(aq)}$$
$$K_{sp} = [Ca^{2+}][F^-]^2$$
$$= [2.0 \times 10^{-4}][2(2.0 \times 10^{-4})]^2$$

Remember, there are two F^- ions for each Ca^{2+} ion.

$$K_{sp} = [2.0 \times 10^{-4}][4.0 \times 10^{-4}]^2$$
$$= 32 \times 10^{-12}$$
$$= 3.2 \times 10^{-11}$$

(2) $K_{sp} = 1.03 \times 10^{-9}$

First calculate the molar mass of $SnCO_3$, using a list of atomic masses.

$$\text{molar mass of } SnCO_3 = (118.7 \text{ g}) + (12.01 \text{ g}) + (3 \times 16.00 \text{ g}) = 178.7 \frac{g}{mol}$$

Next, determine the molarity of the saturated solution of $SnCO_3$, then substitute the values in the solubility product equation.

$$\text{Molarity} = \frac{g/L}{g/mol} = \frac{mol}{L}$$

Substituting in the above equation

$$M = \frac{5.73 \times 10^{-3} \text{ g/L}}{178.7 \text{ g/mol}} = 3.21 \times 10^{-5} \text{ M}$$
$$K_{sp} = [Sn^{2+}][CO_3^{2-}], \ [Sn^{2+}] = [CO_3^{2-}] = 3.21 \times 10^{-5} \text{ M}$$
$$= [3.21 \times 10^{-5}][3.21 \times 10^{-5}]$$
$$= 10.3 \times 10^{-10}$$
$$K_{sp} = 1.03 \times 10^{-9}$$

(3) $K_{sp} = 4.4 \times 10^{-11}$

$$\text{molar mass of } ZnS = 97.44 \frac{g}{mol}$$
$$M = \frac{g/L}{g/mol} = \frac{6.4 \times 10^{-4} \text{ g/L}}{97.44 \text{ g/mol}} = 6.6 \times 10^{-6} \text{ M}$$
$$K_{sp} = [Zn^{2+}][S^{2-}], \ [Zn^{2+}] = [S^{2-}] = 6.6 \times 10^{-6} \text{ M}$$
$$= [6.6 \times 10^{-6}][6.6 \times 10^{-6}]$$
$$= 44 \times 10^{-12}$$
$$K_{sp} = 4.4 \times 10^{-11}$$

11. Acids $HC_2H_3O_2$, acetic acid, is strongest
             HClO, hypochlorous acid, is next
             HCN, hydrocyanic, is weakest

     Salts CaF_2, calcium fluoride, is most soluble
             $PbSO_4$, lead(II) sulfate, is next
             $BaSO_4$, barium sulfate, is least soluble

12. H^+ concentration equals 2.7×10^{-2} M, pH is approximately 1.56. Establish a table as before.

Substance	Initial Conditions	Final Conditions
H_3PO_4	0.10 M	$0.10 - X$
$H_2PO_4^-$	0 M	X
H^+	0 M	X

Substitute the final values into the expression for K_a and assume X has a small value compared to the initial concentration of H_3PO_4.

$$K_a = 7.5 \times 10^{-3} = \frac{[H^+][H_2PO_4^-]}{[H_3PO_4]} = \frac{[X][X]}{[0.10 - X]}$$

$$7.5 \times 10^{-3} = \frac{X^2}{0.10}$$
$$[H^+] = X = \sqrt{7.5 \times 10^{-4}} = 2.7 \times 10^{-2}$$
$$pH = -\log[H^+] = 1.56$$

13. $K_{eq} = 1.98 \times 10^{-2}$, $[H_2] = [I_2] = 4.22 \times 10^{-3}$ M, $[HI] = 2.16 \times 10^{-2}$ M. Quadratic formula yields $[H_2] = [I_2] = 5.88 \times 10^{-3}$ M, $[HI] = 1.83 \times 10^{-2}$ M. The K_{eq} for the reverse reaction is the reciprocal of the K_{eq} for the forward reaction.

$$K_{eq}(\text{reverse}) = \frac{1}{K_{eq}(\text{forward})} = \frac{1}{50.5} = 1.98 \times 10^{-2}$$

The initial concentration of HI is 0.0300 M. Set up a table of values as before

Substance	Initial Conditions	Final Conditions
HI	0.0300 M	$0.0300 - X$
H_2	0 M	$\dfrac{X}{2}$
I_2	0 M	$\dfrac{X}{2}$

One mole of HI decomposes to produce 0.5 mole each of H_2 and I_2.

$$K_{eq} = \frac{[H_2][I_2]}{[HI]^2} = \frac{\left[\dfrac{X}{2}\right]\left[\dfrac{X}{2}\right]}{[0.0300 - X]^2} = \frac{\left[\dfrac{X^2}{4}\right]}{9.00 \times 10^{-4}} = 1.98 \times 10^{-2}$$

Assume X is small compared to 0.0300 M.

Simplifying

$$(1.98 \times 10^{-2})(9.00 \times 10^{-4})(4) = X^2$$

$$X^2 = 7.13 \times 10^{-5}$$

$$X = 8.44 \times 10^{-3}$$

If $X = 8.44 \times 10^{-3}$, then the concentrations of H_2 and I_2 are X/2 or 4.22×10^{-3} M. The concentration of HI is 0.0300 M $-$ 0.00844 M or 0.0216 M.

The value for X is not really small compared to 0.0300 M. Using the quadratic formula, a value for X of 1.17×10^{-2} is obtained rather than 8.44×10^{-3}. In this case the concentrations of H_2 and I_2 would be calculated to be 5.85×10^{-3} M and the concentration of HI would be 0.0183 M.

14. pH = 4.4

The equation is $HC_2H_3O_2 \rightleftarrows H^+ + C_2H_3O_2^-$

$$K_a \frac{[H^+][C_2H_3O_2]}{[HC_2H_3O_2]} = 1.8 \times 10^{-5}$$

0.1 mole sodium acetate furnishes 0.1 mole acetate ion.

$$NaC_2H_3O_2 \rightarrow Na^+ + C_2H_3O_2^-$$

$$\frac{0.10\,mol}{500\,mL} = \frac{0.20\,mol}{1000\,mL} = \frac{0.20\,mol}{L}$$

Substituting into the equation we have

$$\frac{[H^+][0.20]}{[0.40]} = 1.8 \times 10^{-5}$$

$$H^+ = \frac{(1.8 \times 10^{-5})(0.40)}{0.20} = 3.6 \times 10^{-5}\,M$$

$$pH = -\log[H^+] = 4.4$$

The pH can also be calculated from an equation named the Henderson-Hasselbach equation

$$pH = pKa + \log \frac{[salt]}{[acid]}$$

The answer will be the same either way.

15. (1) $NH_4^+ + H_2O \rightleftarrows NH_3 + H_3O^+$

(2) $CO_3^{2-} + 2\,H_2O \rightleftarrows H_2CO_3 + 2\,OH^-$ or $CO_3^{2-} + 2H_2O \rightleftarrows H_2O + CO_2 + 2OH^-$

(3) $CN^- + H_2O \rightleftarrows HCN + OH^-$

(4) $Al^{3+} + 3\,H_2O \rightleftarrows Al(OH)_3 + 3\,H^+$

16. (1) Small pH change. This mixture is a buffer system (weak acid and its conjugate base) so a large pH change is not expected when NaOH is added.

(2) Large pH change. This mixture is made from a strong acid and a neutral salt and so will have the characteristics of a strong acid mixture.

(3) Small pH change. This mixture is a buffer system (weak acid and its conjugate base; the conjugate base was made from the reaction of 1 equivalent of the weak acid with the strong base) so a large pH change is not expected when NaOH is added.

(4) Large pH change. This mixture is made from two strong acids.

WORD SEARCH 2

In the given matrix, find the terms that match the following definitions. Terms may be horizontal, vertical, or on the diagonal. They may also be written forward or backward. Answers are found at the end of chapter 20.

1. A principle that states what happens to an equilibrium system when the conditions are altered.
2. A solution that resists changes in pH.
3. An experimental technique for measuring the volume of one reagent required to react with a measured amount of another reagent.
4. The formation of ions.
5. A dynamic state where two or more opposing processes are taking place at the same time and at the same rate.
6. A homogeneous mixture of two or more substances.
7. A substance whose aqueous solution conducts electricity.
8. Capable of mixing and forming a solution.
9. The H_3O^+ ion.
10. A solution containing 1 mole of solute per liter of solution is 1.0 _____ .
11. A solution containing dissolved solute in equilibrium with undissolved solute.
12. Behaving chemically as either an acid or base.
13. The substance present to the largest extent in a solution.
14. The number of equivalent weights of solute per liter of solution.
15. Incapable of mixing.
16. The substance that is dissolved in a solvent to form a solution.
17. Solution containing a relatively small amount of solute.

A	E	G	Y	Y	F	E	H	Y	K	B	D	I	S	P	D	N	J
G	P	I	S	O	L	U	T	E	O	Q	F	Y	D	I	L	K	Z
X	T	D	N	O	A	I	S	Y	C	I	E	V	L	A	H	P	O
C	K	E	B	G	L	O	M	R	I	X	R	U	N	F	O	J	Z
U	M	T	V	A	E	V	Z	M	A	O	T	E	Q	B	Q	Y	M
S	F	A	M	H	C	R	E	D	I	E	R	E	F	F	U	B	Q
D	N	R	U	S	H	E	R	N	C	S	Q	T	U	W	M	J	P
C	O	U	I	J	A	L	B	P	T	L	C	R	C	G	H	Z	W
N	I	T	N	U	T	B	I	N	E	H	A	I	V	E	E	Y	I
X	T	A	O	W	E	I	C	T	O	L	R	T	B	A	L	O	Q
K	A	S	R	D	L	C	V	M	O	E	W	M	Z	L	V	E	N
V	R	G	D	L	I	S	E	M	T	B	L	N	E	C	E	B	L
O	T	F	Y	X	E	I	I	O	N	I	Z	A	T	I	O	N	R
Y	I	G	H	W	R	M	H	Z	A	X	U	F	A	K	T	T	F
I	T	S	V	A	X	P	U	N	O	I	T	U	L	O	S	G	J
J	P	L	Q	Z	M	U	I	R	B	I	L	I	U	Q	E	E	T
M	C	X	S	A	H	T	A	U	D	F	N	Y	P	R	B	R	W

— 173 —

CHAPTER SEVENTEEN

Oxidation-Reduction

SELECTED CONCEPTS REVISITED

Oxidation numbers allow us to keep track of electrons. You will find it much easier to remember the rules for assigning oxidation numbers if you read carefully section 17.1 in your text. This section explains the basis for the assignment of oxidation numbers and how to use electronegativity to help you determine which element to assign first.

The oxidizing agent causes oxidation but is itself reduced. That is, the oxidizing agent encourages another substance to lose electrons by taking the electrons away. Since the oxidizing agent gained electrons, it is reduced. The reverse argument applies to reducing agents.

When balancing redox reactions, be sure the equation is balanced both atomically and electronically, that is, balance the atoms and the charges. If a redox reaction takes place in aqueous solution, then H_2O is obviously present and can be used as part of a balanced equation. If the reaction is in acidic solution, this means that H^+ ions are also available. In basic solutions, OH^- ions are available to balance the reaction equations.

The higher up in the activity series a species occurs, the more easily it is oxidized. The more easily a species is oxidized, the less likely the product of that oxidation is to be reduced. For example, $K \rightarrow K^+$ is fairly high in the activity series and therefore K is more easily oxidized to K^+ than metals below it in the table (such as $Cu \rightarrow Cu^{2+}$). This means that K^+ is harder to reduce to K than Cu^{2+} is to reduce to Cu.

Reduction occurs at the cathode and oxidation occurs at the anode.

Selected rules for determining oxidation numbers of elements in compounds or ions.
Hydrogen is generally $+1$.
Oxygen is generally -2.
The algebraic sum of the oxidation numbers for all the atoms in a compound is equal to zero.
The algebraic sum of the oxidation numbers for all the atoms in a polyatomic ion is equal to the charge of the ion.

COMMON PITFALLS TO AVOID

Try not to confuse oxidize, reduce, oxidizing agent and reducing agent. The substance being oxidized is losing electrons and is the reducing agent. The species being reduced is gaining electrons and is the oxidizing agent.

Balancing charges does not mean that there must be zero charge on each side of the equation nor that the sum of charges on both sides should equal zero. Balancing charges means that the whatever total charge is on one side of the equation an equal total charge must be present on the other side of the equation.

SELF-EVALUATION SECTION

1. Determine the oxidation number for each indicated element in the following compounds or ions:

    (1) Cr in CrO_3 _____   (9) Cl in ClO_4^- _____

    (2) C in CBr_4 _____   (10) P in P_2O_5 _____

    (3) N in N_2O _____   (11) N in NO_2^- _____

    (4) Mn in MnO_4^- _____   (12) C in $C_6H_{12}O_6$ _____

    (5) As in AsO_4^{3-} _____   (13) N in HNO_3 _____

    (6) Br in BrO_3^- _____   (14) Zn in ZnO_2^{2-} _____

    (7) Mn in $MnO(OH)_2$ _____   (15) C in $C_2O_4^{2-}$ _____

    (8) C in CH_4 _____   (16) S in H_2SO_3 _____

2. In the following reactions, identify the element that is *reduced*, the element that is *oxidized*, the *reducing agent*, and the *oxidizing agent*.

Oxidation Information

   Metallic elements in ionic compounds have a positive charge.
   Elements in the free state have an oxidation number of (1) _____.
   Hydrogen is usually (2) _____.
   Oxygen is usually (3) _____.
   Group 1A metals are always (4) _____.

   (5) $Cr_2O_3 + 3 H_{2(g)} \xrightarrow{\Delta} 2 Cr_{(s)} + 3 H_2O_{(l)}$

   (6) $2 H_{2(g)} + O_{2(g)} \xrightarrow{\Delta} 2 H_2O_{(l)}$

   (7) $2 HNO_{2(aq)} + 2 HI_{(aq)} \rightarrow I_{2(s)} + 2 NO_{(g)} + 2 H_2O_{(l)}$

   (8) $2 Na_{(s)} + H_2O_{(l)} \rightarrow 2 NaOH_{(aq)} + H_{2(g)}$

   (9) $5 NaBr + NaBrO_3 + 3 H_2SO_4 \rightarrow 3 Br_2 + 3 Na_2SO_4 + 3 H_2O$

   Write your answers here.

3. Balance the following unionized and ionic equations using the electron gain-and-loss technique. Place correct coefficients in front of each formula.

(1) $H^+_{(aq)} + Br^-_{(aq)} + SO_4^{2-}_{(aq)} \rightarrow Br_{2(l)} + SO_{2(g)} + H_2O_{(l)}$

(2) $NH_{3(g)} + O_{2(g)} \rightarrow NO_{(g)} + H_2O_{(g)}$

(3) $HNO_{2(aq)} + HI_{(aq)} \rightarrow NO_{(g)} + I_{2(s)} + H_2O_{(l)}$

(4) $Mn^{4+}_{(aq)} + Cl^-_{(aq)} \rightarrow Cl_{2(g)} + Mn^{2+}_{(aq)}$

(5) $ClO_3^-_{(aq)} + SnO_2^{2-}_{(aq)} \rightarrow Cl^-_{(aq)} + SnO_3^{2-}_{(aq)}$

Write your answers here.

4. Balance the following redox equations by the ion-electron method.

(1) $Zn_{(s)} + NO_3^-_{(aq)} \rightarrow Zn^{2+}_{(aq)} + NH_4^+_{(aq)}$ (acidic)

(2) $ClO^- + I^- \rightarrow IO_3^- + Cl^-$ (basic)

Write your answers here.

5. Using the partial list of the activity series of metals, answer the following questions. For any reaction that proceeds write a balanced equation.

$$Mg \rightarrow Mg^{2+} + 2e^-$$
$$Al \rightarrow Al^{3+} + 3e^-$$
$$Zn \rightarrow Zn^{2+} + 2e^-$$
$$Fe \rightarrow Fe^{2+} + 2e^-$$
$$Ni \rightarrow Ni^{2+} + 2e^-$$
$$Sn \rightarrow Sn^{2+} + 2e^-$$
$$Pb \rightarrow Pb^{2+} + 2e^-$$
$$H_2 \rightarrow 2H^+ + 2e^-$$
$$Cu \rightarrow Cu^{2+} + 2e^-$$
$$Ag \rightarrow Ag^+ + e^-$$

(1) Will a chemical reaction occur when a $Pb(NO_3)_2$ solution ($Pb^{2+} + 2\ NO_3^-$) is placed in a *copper* container?

(2) Will a chemical reaction occur when a $ZnCl_2$ solution ($Zn^{2+} + 2\ Cl^-$) is placed in a *magnesium* container?

(3) Will a chemical reaction occur when a HCl solution ($H^+ + Cl^-$) is placed in a *tin* can?

(4) Will a chemical reaction occur when an $AgNO_3$ solution ($Ag^+ + NO_3^-$) is placed in a *aluminum* container?

Do your equations here.

6. Aluminum metal is produced almost entirely by the electrolysis of Al_2O_3. The simplified reactions are

$$Al^{3+} + 3e^- \rightarrow Al$$

$$O^{2-}_{(aq)} + C_{(s)} \rightarrow CO_{(g)} + 2e^-$$

In the sketch below, indicate the direction of ion migration and balance the overall equation.

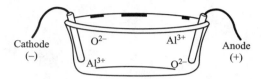

Write your answers here.

7. (1) Write the anode and cathode reactions for the electrolysis of molten $NiBr_2$ with inert electrodes. Then write the balanced overall reaction.

Write your answers here.

(2) Write anode, cathode, and the overall reaction for the electrolysis of molten LiCl with inert electrodes.

Write your answers here.

8. Identify which half-reaction of each pair occurs at the anode and which half-reaction occurs at the cathode.

(1) $Fe^0 \rightarrow Fe^{2+} + 2e^-$
    $Ni_2O_3 + 3\ H_2O + 2e^- \rightarrow 2\ Ni(OH)_2 + 2\ OH^-$

(2) $6\ Fe^{2+} \rightarrow 6\ Fe^{3+} + 6e^-$
    $Cr_2O_7^{2-} + 14\ H^+ + 6e^- \rightarrow 2\ Cr^{3+} + 7\ H_2O$

(3) $2\ H^+ + H_2O_2 + 2e^- \rightarrow 2\ H_2O$
    $2\ I^- \rightarrow I_2 + 2e^-$

Write your answers here.

9. Fill in the blank space with a correct term.

The type of cell that uses chemical reactions to produce electrical energy is called a (1) _____ cell.

The other type of cell, (2) _____ , uses electrical energy to produce a chemical change. A dry cell

flashlight battery is an example of a(an) (3) _____ .

10. The following reactions are involved in the operation of the lead storage battery. Which is the oxidation reaction and which is the reduction reaction? Give the overall balanced cell reaction.

(1) $Pb \rightarrow Pb^{2+} + 2e^-$
(2) $PbO_2 + 4\ H^+ + 2e^- \rightarrow Pb^{2+} + 2\ H_2O$

11. The super iron battery uses K_2FeO_4 according to the equation
    $2\ K_2FeO_4 + 3\ Zn \rightarrow Fe_2O_3 + ZnO + 2\ ZnO_2$
    What are the oxidation numbers for each atom in the above equation?

Challenge Problems

12. Balance the following oxidation-reduction equations. The reaction conditions are either acidic or basic, and H^+ and OH^- plus H_2O should be used accordingly to balance the reactions.

(1) $MnO_4^- + VO^{2+} \rightarrow VO_2^+ + Mn^{2+}$ (acid)

(2) $P_4 \rightarrow PH_3 + H_2PO_2^-$ (basic solution)

(3) $MnO_2 + SO_3^{2-} \rightarrow SO_4^{2-} + Mn(OH)_2$ (basic solution)

(4) $MnO_4^-{}_{(aq)} + H_2O_{2(l)} \rightarrow Mn^{2+}{}_{(aq)} + O_{2(g)}$ (acid)

(5) $Mn^{2+} + HBiO_3 \rightarrow Bi^{3+} + MnO_4^-$ (acid)

(6) $Zn_{(s)} \rightarrow Zn(OH)_4{}^{2-}{}_{(aq)} + H_{2(g)}$ (basic solution)

13. Determine the oxidation number of the indicated element in each of the following compounds.

(1) As in $HAsO_3$ _____ (2) Ti in $Na_2Ti_3O_7$ _____

(3) Pt in $H_2PtCl_6 \cdot 6H_2O$ _____ (4) S in $(NH_4)_2S_2O_8$ _____

(5) S in $Na_2S_2O_3$ _____ (6) Cr in K_2CrO_4 _____

(7) Mo in $(NH_4)_2MoO_4$ _____ (8) Co in $Na_3Co(NO_2)_6$ _____

(9) B in CaB_4O_7 _____ (10) C in $H_2C_2O_4$ _____

RECAP SECTION

The major thrust of Chapter 17 is to give each learner experience at working with chemical equations and manipulating numbers. If you answered the study guide questions without difficulty, you have mastered a valuable chemical tool. You are able to take equations apart, analyze them, and make them work for you. You should realize by now that you have all the skills needed to take a word equation and transform it into a balanced tool to be used for chemical description and calculations. You can handle ions, molecules, or gases in equations and calculations. If you have had difficulty, you should go back over the examples in the text very carefully and then come back to the study guide and the review problems at the end of the chapter.

ANSWERS TO QUESTIONS AND SOLUTIONS TO PROBLEMS

1. (1) +6 (each O is -2 so the oxidation number of the three O's is -6. CrO_3 is neutral so the oxidation numbers sum to zero. Cr must have an oxidation number of +6.)

(2) +4	(3) +1	(4) +7	(5) +5	(6) +5	(7) +4	(8) -4	(9) +7	
(10) +5	(11) +3	(12) 0	(13) +5	(14) +2	(15) +3	(16) +4		

2. (1) 0 (2) +1 (3) -2 (4) +1

(5) Hydrogen gas is oxidized and is the reducing agent. Chromium is reduced and is the oxidizing agent.

$$Cr_2O_{3(s)} + 3\ H_{2(g)} \xrightarrow{\Delta} 2\ Cr_{(s)} + 3\ H_2O_{(l)}$$

First, assign oxidation numbers. $O = -2$ and $H = +1$. Hydrogen gas = 0. Chromium metal = 0.

$$\underset{+3\ -2}{Cr_2O_3} + \underset{0}{3\ H_2} \rightarrow \underset{0}{2\ Cr} + \underset{+1\ -2}{3\ H_2O}$$

What is Cr in Cr_2O_3? If $O = -2$, then each Cr will have to be $+3$ in order to balance the total amount of negative oxidation value of 3 oxygen. Which element has been oxidized (lost electrons)? Hydrogen has changed from 0 to $+1$, so it has been oxidized. Cr has changed from $+3$ to 0, so it has been reduced.

(6) Hydrogen gas oxidized and is the reducing agent. Oxygen gas is reduced and is the oxidizing agent.

$$\underset{0}{2\ H_{2(g)}} + \underset{0}{O_{2(g)}} \xrightarrow{\Delta} \underset{+1\ -2}{2\ H_2O_{(g)}}$$

Each hydrogen atom has lost an electron, and each oxygen atom has gained two electrons.

(7) Iodine is oxidized and is the reducing agent. Nitrogen is reduced and is the oxidizing agent.

$$\underset{+1\ +3\ -2}{2\ HNO_{2(aq)}} + \underset{+1\ -1}{2\ HI_{(aq)}} \rightarrow \underset{0}{I_{2(s)}} + \underset{+2\ -2}{2\ NO_{(g)}} + \underset{+1\ -2}{2\ H_2O_{(l)}}$$

Iodine changes from -1 to 0, which is a loss of one electron (oxidation). Nitrogen has gained one electron in changing from $+3$ to $+2$.

(8) Sodium is oxidized and hydrogen is reduced. Sodium is the reducing agent, and hydrogen is the oxidizing agent.

$$\underset{0}{2\ Na} + \underset{+1\ -2}{H_2O} \rightarrow \underset{+1\ -2\ +1}{2\ NaOH} + \underset{0}{H_2}$$

Sodium changes from 0 to $+1$, an oxidation process. Na is the reducing agent. Hydrogen changes from $+1$ to 0 for one of the products, hydrogen gas. Notice that hydrogen is also present in NaOH at the same oxidation state as in H_2O. Hydrogen is reduced and is the oxidizing agent.

(9) Bromine is both oxidized and reduced.

$$\underset{+1\ -1}{5\ NaBr} + \underset{+1\ +5\ -2}{NaBrO_3} + \underset{+1\ +6\ -2}{3\ H_2SO_4} \rightarrow \underset{0}{3\ Br_2} + \underset{+1\ +6\ -2}{3\ Na_2SO_4} + 3\ H_2O$$

The only atoms changing oxidation number are Br. Five Br atoms change from -1 to 0. This is oxidation, a loss of electrons. These Br atoms are the reducing agents, The other Br atom changes from $+5$ to 0, which is a gain of electrons, or reduction. This Br atom is the oxidizing agent.

3. (1) $4 H^+_{(aq)} + 2 Br^-_{(aq)} + SO_4^{2-}_{(aq)} \rightarrow Br_{2(l)} + SO_{2(g)} + 2 H_2O_{(l)}$

Assign oxidation numbers and write each half-reaction for oxidation and reduction.

$$H^+_{(aq)} + Br^-_{(aq)} + SO_4^{2-}_{(aq)} \rightarrow Br_{2(l)} + SO_{2(g)} + 2 H_2O_{(l)}$$
$$+1 \qquad -1 \qquad +6 -2 \qquad 0 \qquad +4 -2 \qquad +2 -2$$

Oxidation $2 Br^- \rightarrow Br_2 + 2e^-$

Reduction $S^{6+} + 2e^- \rightarrow S^{4+}$

Electron gain and loss is balanced. Insert a 2 in front of Br^-.

$$H^+_{(aq)} + 2 Br^-_{(aq)} + SO_4^{2-}_{(aq)} \rightarrow Br_{2(l)} + SO_{2(g)} + H_2O_{(l)}$$

Check for balance of atoms. H and O are still out of balance. We need four H^+ on the left side and two H_2O molecules on the right.

$$4 H^+_{(aq)} + 2 Br^-_{(aq)} + SO_4^{2-}_{(aq)} \rightarrow Br_{2(l)} + SO_{2(g)} + 2 H_2O_{(l)}$$

Do a check on electrical charges.

$$(4+) + (2-) + (2-) \qquad 0$$
$$\text{left side} \qquad\qquad \text{right side}$$

(2) $4 NH_{3(g)} + 5 O_{2(g)} \rightarrow 4 NO_{(g)} + 6 H_2O_{(g)}$

Assign oxidation numbers and write each half-reaction for oxidation and reduction.

$$NH_{3(g)} + O_{2(g)} \rightarrow 4 NO_{(g)} + H_2O_{(g)}$$
$$-3 +1 \qquad 0 \qquad +2 -2 \qquad +1 -2$$

This equation has a small wrinkle to it. Oxygen is reduced from 0 to -2 and shows up in two molecules on the product side

Oxidation $N^{3-} \rightarrow N^{2+} + 5e^-$

Reduction $O_2^0 + 4e^- \rightarrow 2 O^{2-}$ (from NO and H_2O)

To balance a gain of $4e^-$ and a loss of $5e^-$, we must use 20 as the lowest common denominator.

$$4 N^{3-} \rightarrow 4 N^{2+} + 20e^-$$
$$5 O_2 + 20e^- \rightarrow 10 O^{2-}$$

Put coefficients back in the equation. Remember that we can only have four NO molecules and that the rest of the oxygen on the product side is in water.

$$4\ NH_{3(g)} + 5\ O_{2(g)} \rightarrow 4\ NO_{(g)} + H_2O_{(g)}$$

To balance out the hydrogen, place a 6 in front of the water. The oxygen is now also balanced.

$$4\ NH_{3(g)} + 5\ O_{2(g)} \rightarrow 4\ NO_{(g)} + 6\ H_2O_{(g)}$$

(3) $2\ HNO_{2(aq)} + 2\ HI_{(aq)} \rightarrow 2\ NO_{(g)} + I_{2(s)} + 2\ H_2O_{(l)}$

Assign oxidation numbers and write each half-reaction for oxidation and reduction.

$$HNO_2 + HI \rightarrow NO + I_2 + H_2O$$
$${+1+3-2}\ \ \ {+1-1}\ \ \ {+2-2}\ \ \ 0\ \ \ {+1-2}$$

Oxidation	$I^- \rightarrow I_2$	
balance	$2\ I^- \rightarrow I_2 + 2e^-$	(I^- loses 1 electron)
Reduction	$N^{3+} \rightarrow N^{2+}$	
balance	$N^{3+} + 1e^- \rightarrow N^{2+}$	(N^{3+} gains 1 electron)

Multiply the reduction equation by 2 to balance the electron gain and loss.

$$2\ I^- \rightarrow I_2 + 2e^-$$
$$2\ N^{3+} + 2e^- \rightarrow 2\ N^{2+}$$

Place each half-reaction along with the coefficient back in the original equation.

$$2\ HNO_{2(aq)} + 2\ HI_{(aq)} \rightarrow 2\ NO_{(g)} + I_{2(s)} + 2\ H_2O_{(l)}$$

Balance the remaining H and O atoms.

$$2\ HNO_{2(aq)} + 2\ HI_{(aq)} \rightarrow 2\ NO_{(g)} + I_{2(s)} + 2\ H_2O_{(l)}$$

(4) $Mn^{4+}_{(aq)} + 2\ Cl^-_{(aq)} \rightarrow Cl_{2(g)} + Mn^{2+}_{(aq)}$

This equation should be fairly simple to balance.

$$Mn^{4+}_{(aq)} + Cl^-_{(aq)} \rightarrow Cl_{2(g)} + Mn^{2+}_{(aq)}$$

Cl^- is oxidized to free elemental chlorine, and Mn^{4+} is reduced to Mn^{2+}.

Oxidation	$2\ Cl^-_{(aq)} \rightarrow Cl_{2(g)} + 2e^-$	(1 electron lost per atom, 2 electrons lost per molecule)
Reduction	$Mn^{4+}_{(aq)} + 2e^- \rightarrow Mn^{2+}_{(aq)}$	(2 electrons gained per atom)

$$Mn^{4+}{}_{(aq)} + 2\ Cl^-{}_{(aq)} \rightarrow Cl_{2(g)} + Mn^{2+}{}_{(aq)}$$

The equation is now balanced. Check the electrical charges for the last detail.

$$
\begin{array}{cc}
(4+) + (2-) & 2+ \\
\text{left side} & \text{right side}
\end{array}
$$

The equation must have the same electrical charge on both sides as well as the same number of atoms of each element.

(5) $ClO_3^-{}_{(aq)} + 3\ SnO_2{}^{2-}{}_{(aq)} \rightarrow Cl^-{}_{(aq)} + 3\ SnO_3{}^{2-}{}_{(aq)}$

Assign oxidation numbers and write each half-reaction for oxidation and reduction.

$$
\begin{array}{cccc}
ClO_3^-{}_{(aq)} + & SnO_2{}^{2-}{}_{(aq)} \rightarrow & Cl^-{}_{(aq)} + & 3\ SnO_3{}^{2-}{}_{(aq)} \\
{\scriptstyle +5\ -2} & {\scriptstyle +2\ -2} & {\scriptstyle -1} & {\scriptstyle +4\ -2}
\end{array}
$$

Oxidation $Sn^{2+}{}_{(aq)} \rightarrow Sn^{4+}{}_{(aq)} + 2e^-$ (2e$^-$ loss per atom)

Reduction $Cl^{5+}{}_{(aq)} + 6e^- \rightarrow Cl^-{}_{(aq)}$ (6e$^-$ gain per atom)

Balance the electron gain and loss first.

$$3\ Sn^{2+}{}_{(aq)} \rightarrow 3\ Sn^{4+}{}_{(aq)} + 6e^-$$
$$Cl^{5+}{}_{(aq)} + 6e^- \rightarrow Cl^-{}_{(aq)}$$

Place each coefficient back in the original equation.

$$ClO_3^-{}_{(aq)} + 3\ SnO_2{}^{2-}{}_{(aq)} \rightarrow Cl^-{}_{(aq)} + 3\ SnO_3{}^{2-}{}_{(aq)}$$

Check whether the atoms of oxygen balance. Do the electrical charges balance also?

$$
\begin{array}{cc}
(1-) + (6-) & (-1) + (6-) \\
\text{left side} & \text{right side}
\end{array}
$$

4. (1) $4\ Zn + NO_3^- + 10\ H^+ \rightarrow 4\ Zn^{2+} + NH_4^+ + 3\ H_2O$

To begin, write the two half-reactions containing the elements being oxidized and reduced.

Oxidation $Zn \rightarrow Zn^{2+}$

Reduction $NO_3^- \rightarrow NH_4^+$

The Zn and N are balanced on each side, so now balance the H and O. Remember acid conditions.

$$10\ H^+ + NO_3^- \rightarrow NH_4^+ + 3\ H_2O$$

Now balance the electrons in the NO_3^- half-reaction.

$$8e^- + 10\ H^+ + NO_3^- \rightarrow NH_4^+ + 3\ H_2O$$

— 185 —

We now have 8e$^-$ available for the oxidation half-reaction of Zn solid. This means we can use four atoms of Zn for each NO_3^-.

$$4 \, Zn \rightarrow Zn^{2+} + 8e^-$$

Adding the two half-reactions together we have

$$4 \, Zn + NO_3^- + 10 \, H^+ \rightarrow 4 \, Zn^{2+} + NH_4^+ + 3 \, H_2O$$

(2) $3 \, ClO^- + I^- \rightarrow 3 \, Cl^- + IO_3^-$

Write the two half-reactions

Oxidation $I^- \rightarrow IO_3^-$

Reduction $ClO^- \rightarrow Cl^-$

The I and Cl are balanced. Now proceed with the H and O using basic conditions.

To balance O and H in the oxidation equation, add 3 H_2O to the left and 6H^+ to the right side of the equation.

$$I^- + 3 \, H_2O \rightarrow IO_3^- + 6 \, H^+$$

Add 6 OH$^-$ to each side

$$6 \, OH^- + I^- + 3 \, H_2O \rightarrow IO_3^- + 6 \, H^+ + 6 \, OH^-$$

Combine $H^+ + OH^- \rightarrow H_2O$

$$6 \, OH^- + I^- + 3 \, H_2O \rightarrow IO_3^- + 6 \, H_2O$$

Rewrite canceling H_2O on each side

$$6 \, OH^- + I^- \rightarrow IO_3^- + 3 \, H_2O$$

To balance O and H in the reduction equation, add 1 H_2O to the right side of the equation and 2 H^+ to the left side.

$$2 \, H^+ + ClO^- \rightarrow Cl^- + H_2O$$

Add 2 OH$^-$ to each side.

$$2 \, OH^- + 2 \, H^+ + ClO^- \rightarrow Cl^- + H_2O + 2 \, OH^-$$

Combine $H^+ + OH^- \rightarrow H_2O$.

$$2 \, H_2O + ClO^- \rightarrow Cl^- + H_2O + 2 \, OH^-$$

Rewrite canceling H_2O on each side.

$$H_2O + ClO^- \rightarrow Cl^- + 2 \, OH^-$$

Balance each half-reaction electrically with electrons.

$$6 \, OH^- + I^- \rightarrow IO_3^- + 3 \, H_2O + 6e^-$$
$$H_2O + ClO^- + 2e^- \rightarrow Cl^- + 2 \, OH^-$$

Equalize loss and gain of electrons. Multiply reduction reaction by 3.

$$6\,OH^- + I^- \rightarrow IO_3^- + 3\,H_2O + 6e^-$$
$$3\,H_2O + 3\,ClO^- + 6e^- \rightarrow 3\,Cl^- + 6\,OH^-$$

Add the two half-reactions together, canceling the $6e^-$, $3\,H_2O$ and $6\,OH^-$ from each side of the equation.

$$3\,ClO^- + I^- \rightarrow IO_3^- + 3\,Cl^-$$

Check: each side of the equation has a charge of -4 and contains the same number of atoms of each element.

5. (1) No reaction. Copper is below lead on the series and therefore will not replace lead ions from solution.

   (2) Reaction. Magnesium is above zinc on the series and will therefore replace zinc ions from solution.

$$Mg + ZnCl_2 \rightarrow Zn + MgCl_2$$

   (3) Reaction. Tin is above hydrogen on the series and will therefore replace hydrogen ions from solution. Reaction will occur.

$$Sn + 2\,HCl \rightarrow SnCl_2 + H_2$$

   (4) Reaction. Aluminum is above silver on the series and will replace silver ions from solution.

$$Al + 3\,AgNO_3 \rightarrow Al(NO_3)_3 + 3\,Ag$$

6. The Al^{3+} ions migrate toward the cathode, and the O^{2-} ions migrate toward the anode. To balance the two half-reactions, balance the electron gain and loss.

$$Al^{3+} + 3e^- \rightarrow Al$$
$$O^{2-} + C \rightarrow CO + 2e^-$$

If you multiply the Al^{3+} half-reaction by 2 and the O^{2-} half-reaction by 3, the electrons will be balanced.

$$2\,Al^{3+} + 6e^- \rightarrow 2\,Al$$
$$3\,O^{2-} + 3\,C \rightarrow 3\,CO + 6e^-$$

Added together: $2\,Al^{3+} + 3\,O^{2-} + 3\,C \rightarrow 2\,Al + 3\,CO$

7. (1) Molten $NiBr_2$ will exist as the following ions $NiBr_2 \xrightarrow{heat} Ni^{2+} + 2\,Br^-$. Oxidation, which is a loss of electrons, occurs at the anode. Br^- is the ion capable of losing electrons. The Ni^{2+} ion needs $2e^-$ (reduction) to become a Ni atom.

Therefore, anode reaction $2\,Br^- \rightarrow Br_2 + 2e^-$

The Ni^{2+} ion needs $2e^-$ (reduction) to become a Ni atom.

Cathode reaction $\quad Ni^{2+} + 2e^- \rightarrow Ni^0$

Overall reaction $\quad Ni^{2+} + 2\,Br^- \rightarrow Ni^0 + Br_2$

(2)   LiCl will exist as Li^+ and Cl^- ions in the molten state. Oxidation at the anode will involve the Cl^- ion, but 2 Cl^- ions are needed to balance the equation since Cl_2 is produced.

Anode $2\,Cl^- \rightarrow Cl_2 + 2e^-$

Reduction at the cathode will use the $2e^-$ available to reduce 2 Li^+ ions to lithium atoms.

Cathode $2e^- + 2\,Li^+ \rightarrow 2\,Li^0$

Overall $2\,Li^+ + 2\,Cl^- \rightarrow 2\,Li^0 + Cl_2$

8. (1) Iron goes from the elemental state to 2+. This involves a loss of two electrons and takes place at the anode. Nickel is reduced from the 3+ state to 2+ state at the cathode.

(2) Fe^{2+} loses $1e^-$ to reach the Fe^{3+} oxidation state. Loss of e^- is oxidation, which occurs at the anode. Cr^{6+} goes from 6+ to 3+, which is a gain of electrons or reduction. This occurs at the cathode.

(3) In hydrogen peroxide (H_2O_2), oxygen is in a 1− oxidation state whereas in water oxygen is its usual 2−. This is reduction and occurs at the cathode. I^- loses electrons to reach the elemental state or zero oxidation state, I_2. This is oxidation or the anode reaction.

These examples may be somewhat difficult. Look at them carefully and be sure you are clear on how to determine oxidation states of various atom. Refer to the text if necessary.

9. (1) voltaic   (2) electrolytic   (3) voltaic

10. Reaction (1) is oxidation, reaction (2) is reduction. Overall reaction:

$Pb + PbO_2 + 4\,H^+ \rightarrow 2\,Pb^{2+} + 2\,H_2O$

11.

$2\,K_2FeO_4 + 3\,Zn \rightarrow Fe_2O_3 + ZnO + 2\,K_2ZnO_2$
$\quad$ +1 +6 −2 $\quad$ 0 $\quad$ +3 −2 $\quad$ +2 −2 $\quad$ +1 +2 −2

K_2FeO_4: each K is +1, each O is −2 so $(2)(+1) + (4)(+2) + x = 0$; $x(Fe) = +6$
Zn in its elemental state has an oxidation number of 0
Fe_2O_3: each O is −2 so $2x + 3(-2) = 0$; $x(Fe) = +3$
ZnO: each O is −2 so $x + (-2) = 0$; $x(Zn) = +2$
K_2ZnO_2: each K is +1, each O is −2 so $2(+1) + x + 2(-2) = 0$; $x(Zn) = +2$

12. (1) $MnO_4^- + 5\,VO^{2+} + H_2O \rightarrow 5\,VO_2^+ + 2\,H^+ + Mn^{2+}$

(2) $P_4 + 3\,H_2O + 3\,OH^- \rightarrow PH_3 + 3\,H_2PO_2^-$

(3) $MnO_2 + SO_3^{2-} + H_2O \rightarrow Mn(OH)_2 + SO_4^{2-}$

(4) $5\,H_2O_2 + 2\,MnO_4^- + 6\,H^+ \rightarrow 2\,Mn^{2+} + 5\,O_2 + 8\,H_2O$

(5) $2\,Mn^{2+} + 5\,HBiO_3 + 9\,H^+ \rightarrow 2\,MnO_4^- + 5\,Bi^{3+} + 7\,H_2O$

(6) $Zn_{(s)} + 2\,OH^-_{(aq)} + 2\,H_2O_{(l)} \rightarrow Zn(OH)_4^{2-}{}_{(aq)} + H_{2(g)}$

13. (1) +5   (2) +4   (3) +4   (4) +7   (5) +2
(6) +6   (7) +6   (8) +3   (9) +3   (10) +3

CHAPTER EIGHTEEN

Nuclear Chemistry

SELECTED CONCEPTS REVISITED

Nuclear reactions involve the nucleus of the atom and therefore the product of a nuclear reaction can be, and often is, a different element. In all the previous reactions you have seen, only the electrons were shifted and the elements involved in the reactant were still part of the product. For a nuclear reaction, the protons, neutrons and electrons may all be involved and therefore the product may involve a different element.

Balancing nuclear reactions involves balancing not the elements but the atomic numbers and mass numbers. Therefore it is important that all reactants and products be written in isotopic notation. The mass numbers are balanced independently of the atomic numbers. The elements or particles are assigned based on the atomic number of the particle. Table 18.1 in your text shows the isotopic notation for radioactive and subatomic particles.

Nuclear fission is the splitting of a nuclide using a neutron.
Nuclear fusion is the joining of two nuclides.
Both fission and fusion release large quantities of energy.

COMMON PITFALLS TO AVOID

A mass of radioactive material does not disappear after two half-lives! The half-life is the time it takes for the radioactive sample to be reduced (decay) to half its current mass. For example, if the half-life of 48 g of a radioactive sample were 1 day, then after 1 day, only 24 g would be left. But the half-life for the 24 g quantity is also one day so at the end of the second day (after a total of two half-lives) 12 g would still be present.

SELF-EVALUATION SECTION

1. Consider the nuclide $^{80}_{27}X$.

    (1) Fill in the blanks with one of the following choices.
         element symbol, atomic number, mass number

    (a) _____ ⟵ ——— $^{80}_{27}X$ ———⟶ (c) _____
    (b) _____ ⟵ ———

(2)   What are the three principal rays or particles coming from the nucleus of a radioactive nuclide?
      (a) _____, (b) _____, and (c) _____.

(3)   Assume $^{80}_{27}X$ can decay by all three forms of radioactivity. Write the symbol of the resulting nuclide. Use a different letter to indicate a different element.

      (a) after alpha decay _____

      (b) after beta decay _____

      (ci) after gamma radiation _____

(4)   A 16.0 gram sample of $^{-223}_{88}Ra$ takes 11.2 days to decay to 8.0 grams. The 11.2 days is called the _____ of the nuclide.

(5)   How many days would it take the original 16.0 gram sample of $^{-223}_{88}Ra$ to decay to 0.5 grams?

2.   Fill in the blanks

	alpha	beta	gamma
(1) symbol	____	____	____
(2) composition	____	____	____

3.   For each of the following characteristics, place the three common types of radiation in the correct order.

	Most		Least
(1) Ionizing power	____	____	____
(2) Heaviest to lightest	____	____	____
(3) Velocity of radiation	____	____	____
(4) Electrical charge (+)	____	____	____ (-)
(5) Penetrating ability of radiation	____	____	____

4.   Fill in the missing atomic number, atomic mass, or missing symbol for the following reactions. You may need to use a periodic table.

(1) $^{238}_{92}U \rightarrow ^{4}_{2}He +$ _____

(2)   _____ $\rightarrow ^{214}_{83}Bi + ^{0}_{-1}e$

(3) $^{210}_{84}Po \rightarrow ^{206}_{82}Pb +$ _____

(4) $^{209}_{83}Bi +$ _____ $\rightarrow ^{209}_{83}Po + ^{1}_{0}n$

(5) $^{35}_{17}Cl + ^{1}_{0}n \rightarrow$ _____ $+ ^{1}_{1}H$

(6)   _____ $+ ^{1}_{0}n \rightarrow ^{24}_{11}Na + ^{4}_{2}He$

(7) $^{14}_{7}N +$ _____ $\rightarrow ^{14}_{6}C + ^{1}_{1}H$

(8) $^{14}_{7}N + ^{1}_{0}n \rightarrow ^{12}_{6}C +$ _____

(9) $^{60}_{28}Ni + ^{1}_{1}H \rightarrow ^{57}_{27}Co +$ _____

(10) $^{7}_{3}Li + ^{1}_{1}H \rightarrow$ _____ $+ ^{1}_{0}n$

5. Identify the following reactions as nuclear fission or nuclear fusion reactions.

(1) $_{1}^{2}H + _{1}^{2}H \rightarrow _{2}^{3}He + _{0}^{1}n$

(2) $_{92}^{238}U + _{0}^{1}n \rightarrow _{56}^{144}Ba + _{36}^{90}Kr + 2 _{0}^{1}n$

(3) $_{3}^{7}Li + _{1}^{1}H \rightarrow 2 _{2}^{4}He$

(4) $_{1}^{2}H + _{1}^{2}H \rightarrow _{1}^{3}H + _{1}^{1}H$

(5) $_{0}^{1}n + _{92}^{238}U \rightarrow _{54}^{144}Xe + _{38}^{90}Sr + 2 _{0}^{1}n$

(6) $_{1}^{3}H + _{1}^{2}H \rightarrow _{2}^{4}He + _{0}^{1}n$

6. Fill in the blank space or circle the correct response.

High levels of radiation, especially gamma or X rays, are termed (1)_____ radiation. If the dosage is high enough, (2) _____ can occur within several days. The effects of radiation appear to be localized in the (3) _____ of cells. Rapidly growing and dividing cells are (4) most/least susceptible to damage. Long-term or protracted exposure to (5) high/low levels of radiation can lead to health problems at some later time in a person's life. Presently there is concern about the effect that strontium-90, which is chemically similar to the element (6) _____ , has on blood cells manufactured in bone marrow. An accumulation of Sr-90 may lead to increased incidence of bone cancer and leukemia. Radiation damage to the nucleus of a cell can affect future generations of a particular species by giving rise to genetic (7) _____ . Such events occur when the radiation damages the genetic material – a molecule called (8) _____ – but not severely enough to prevent reproduction.

7. A radioactive nuclide undergoes the following disintegration series:
α, β, β, α, α, α, α, α, γ, α, β, β, α, β, β, α, α, γ, α.
(1) By how much has the nuclide lost/gained in atomic number?
(2) By how much has the nuclide lost/gained in mass number?

8. Match the names of scientists associated with nuclear chemistry with the appropriate descriptive phrase.

(1) Becquerel   a. Discovered alpha and beta rays
(2) Marie Curie   b. Developed cyclotron
(3) Rutherford   c. Coined word "radioactivity"
(4) E. O. Lawrence   d. Found that uranium salts emit rays
(5) Hahn and Strassmann   e. Reported first instance of nuclear fission

9. Fill in the blank space or circle the correct response.

A magnetic field affects the three principal radioactive rays differently. The beta particle, being (1) positively/negatively charged, will be attracted (2) toward/away from the positive plate. The gamma ray, which has a mass of (3) _____ and an electrical charge of (4) _____ , (5) will/will not be attracted by the magnetic field. The alpha ray, which has a charge of (6) _____ , will be deflected (7) _____ the negative plate.

10. A piece of a wooden tool found at a Northwest Indian fishing village site has been determined to be approximately 10,000 years old. The ^{14}C in the wood has undergone approximately how many half lives of decay? ^{14}C has a half-life of 5668 years.

Write your answer here.

11. The half life for $^{32}_{15}$P, a common biological radionuclide, is 14.3 days. Starting with 1500 micrograms of $^{32}_{15}$P, how much will you have left after 100 days?

Do your calculations here.

12. Fill in the blank space or circle the correct response.

The mass of an atomic nucleus is (1) more/less than the sum of the masses of the particles that make up the nucleus. The difference in mass is known as the (2) _____ . The energy equivalent to this mass (using the equation $E = mc^2$) is called the (3) _____ of the nucleus. This amount of energy would be required to (4) put together/pull apart the particles of a particular nucleus. The higher the binding energy, the (5) more/ less stable the nucleus is. In both nuclear fission and fusion reactions, the products have less mass than the reactants. The resultant mass losses are accounted for in the very large quantities of energy that are released.

Challenge Problems

13. Calculate the binding energy for $^{32}_{16}$S which occurs naturally at an abundance of 95%. The mass of one atom of $^{32}_{16}$S is known to be 31.97207 amu (atomic mass units). The masses of the elemental particles are as follows:
proton = 1.007277 amu   neutron = 1.008665 amu
electron = 0.0005486 amu   1 amu = 1.49×10^{-10} J

Do your calculations here.

14. The binding energy of one $_3^7$Li atom is 6.258×10^{-12} J. Using the masses of the elemental particles listed in problem 12, calculate the actual mass (in amu) of $_3^7$Li atom.

$$1 \text{ amu} = 1.49 \times 10^{-10} \text{ J}$$

Do your calculations here.

RECAP SECTION

Chapter 18 is a fascinating chapter about a subject that influences us all. We are constantly bombarded by cosmic radiation, nuclear power plants pose problems of disposal of highly radioactive wastes, and we may sometime come into contact with radioactive isotopes during medical treatment. Radioactivity is not something to fear, but it is something to be discussed with knowledge and respect. Every citizen should know the essential details concerning radioactivity and its current and potential uses.

ANSWERS TO QUESTIONS AND SOLUTIONS TO PROBLEMS

1. (1) (a) mass number   (b) atomic number   (c) element symbol

   (2) (a) alpha   (b) beta   (c) gamma

   (3)   (a) $_{25}^{76}$Z (the mass number is 4 less, the atomic number is 2 less and the nuclide therefore has different elemental symbol)

         (b) $_{28}^{80}$Y (the mass number does not change, the atomic number increases by 1 and the nuclide therefore has different elemental symbol)

         (c) $_{27}^{80}$X (gamma radiation has no mass and therefore there is no change in symbol)

   (4) half-life

   (5) 56 days (5 half-lives: 16.0 --- 8.0 --- 4.0 --- 2.0 ---1.0 ---0.5)

2. (1) α $_2^4$He β $_{-1}^0$e γ
   (2) identical to He$_{2+}$ identical to an electron   photons of light

3. (1) γ, β, α (2) α, β, γ (3) γ, β, α (4) α, γ, β (5) γ, β, α

4. (1)   Missing species would be $_{90}^{234}$Th.

   (2)   Missing species would be $_{82}^{214}$Pb.

(3)   Missing species would be an alpha particle, $_2^4\text{He}$

(4)   Missing species would be an deuterium atom, $_1^2\text{H}$.

(5)   Missing species would be $_{16}^{35}\text{S}$.   (6)   Missing species would be $_{13}^{27}\text{Al}$.

(7)   Missing species would be $_0^1\text{n}$.   (8)   Missing species would be $_1^3\text{H}$.

(9)   Missing species would be $_2^4\text{He}$.   (10)   Missing species would be $_4^7\text{Be}$.

5.   (1)   Nuclear fusion – two light atoms combine to form a heavier nucleus.
      (2)   Nuclear fission – heavy, unstable nucleus splits into two smaller fragments under bombardment with neutrons.
      (3)   Nuclear fusion   (4)   Nuclear fusion   (5)   Nuclear fission.

6.   (1)   acute   (2)   death   (3)   nucleus   (4)   most
      (5)   low   (6)   calcium   (7)   mutations   (8)   DNA

7.   There are a total of 11 α decays and 6 β decays; γ-radiation will have no effect. 11 alpha decays results in a loss of 44 in mass number and 22 in atomic number. 6 beta decays results in no change in atomic mass and a gain of 6 in atomic number.
      (1)   Atomic number will go down by 16.
      (2)   Atomic mass will be reduced by 44.

8.   (1) d   (2) c   (3) a   (4) b   (5) e

9.   (1) negatively   (2)   toward   (3)   none   (4)   zero
      (5) will not   (6) 2^+ (7)   toward

10.   Approximately 2.

      The half-life for $_6^{14}\text{C}$ is 5668 years

11.   12 μg
      100 days is 7 half-lives. After this period of time 0.78% of the original material will be left.
          $0.0078 \times 1500 \ \mu\text{g} = 12 \ \mu\text{g}$

12.   (1)   less   (2)   mass defect   (3)   binding energy   (4)   pull apart
      (5)   more

13.   Mass defect is equal to 0.29178 amu which is equal to 4.3475×10^{-11} J/atom of $_{16}^{32}\text{S}$.

14.   7.016 amu
      The calculated mass of the elemental particles in a $_3^7\text{Li}$ atom gives a value of 7.058137 amu.
      The binding energy is equal to 4.2×10^{-2} amu.

CHAPTER NINETEEN

Introduction to Organic Chemistry

SELECTED CONCEPTS REVISITED

Learning the structure of the functional groups is crucial to your knowledge of organic chemistry. The nomenclature, physical properties and chemical properties of organic compounds are all based on the functional groups present in the molecule. Being able to recognize the different functional groups in a molecule therefore gives you a large insight into the potential reactivity of the compound.

Although it may appear as though there are a different set of rules for naming each type of functional group on compounds, the underlying premise and process is actually fairly simple. A worked example given on the next page summarizes the general nomenclature process.

A popular question for homework or exams is asking a student to correct an incorrectly named organic compound. Should you be asked to correct the name of a compound, first draw the compound from the name given, then ignore that potentially incorrect name. The most common errors to check for include checking to see that the longest chain is correctly found, that the branches/substituents are numbered from the correct end of the chain, and that the branches/substituents are listed alphabetically in the name.

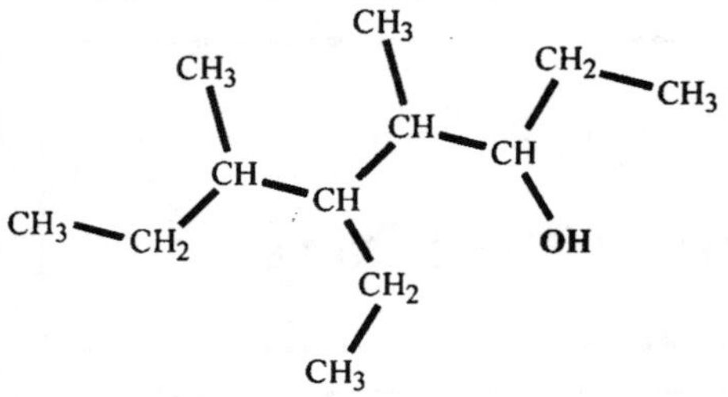

The molecule at left is the same as that at right. The only difference is that the CH_3, CH_2 and CH groups are not shown at right, only their connecting bonds. Therefore, the end of each line and each intersection actually has a carbon atom present.

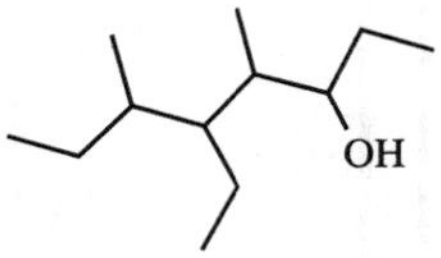

What to look for:	Impact on name:	Example
find the functional group(s) in the molecule such as alkene or ketone groups. E.g., alcohol	determines the ending of the name, such as -ene, -one. E.g., "ol".	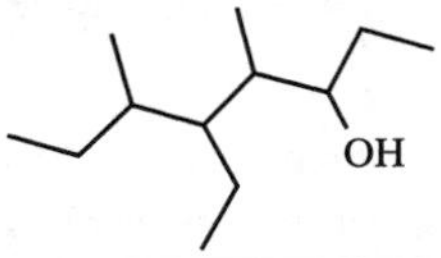
find the longest straight chain that includes the functional group. E.g., eight carbons	determines the root of the main part of the name that gets placed before the ending found above. E.g., **octanol**	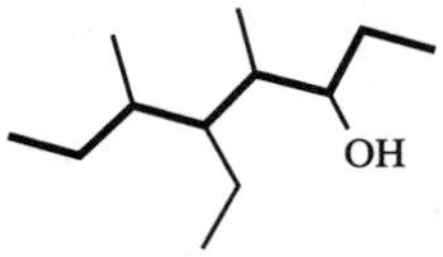
number from the end of the straight chain that gives the functional group the lowest possible number (if you have an alkane, number such that the first branch has the lowest possible number). E.g., number from right to left	determines the number that immediately precedes the name you have thus far – the number indicates the position of the FG on the main chain. E.g., **3-octanol**	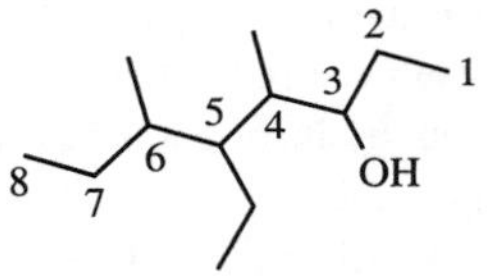
find any branches or substituents; number and name them. E.g., branches are on carbons 4,5,6 of the main chain.	E.g., a methyl group on carbon 4 and another on carbon 6 and an ethyl group on carbon 5 will be named as follows: **4-methyl** **6-methyl** **5-ethyl**	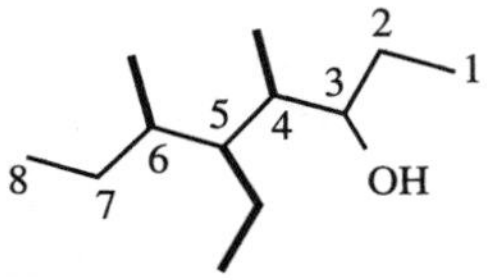
organize the branches in alphabetical order (keeping the numbers assigned to them), then combine any similar branches and prefix the branch name with di, tri, etc. as appropriate.	E.g., 5-ethyl-4-methyl-6-methyl gets shortened to 5-ethyl-**4,6-di**methyl-	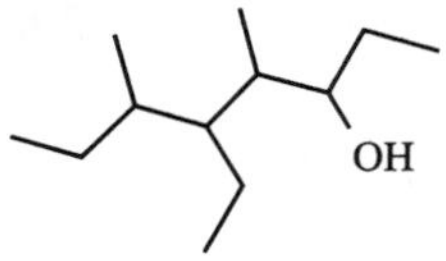
Put the whole name together with the branches first.	E.g., 5-ethyl-4,6-dimethyl-3-octanol	

COMMON PITFALLS TO AVOID

Carbon atoms *cannot* have 5 bonds! Carbon likes having four bonds (satisfies the octet rule). A carbon with only three bonds to it must be charged.

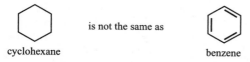

cyclohexane   is not the same as   benzene

If you see an OH on a molecule, please take a closer look to verify whether it is an alcohol (ROH) or a carboxylic acid (RCOOH).

If you see an OR group on a molecule, please take a closer look to verify whether it is an ether (ROR'), an ester (RCOOR'), or a ketone (RCOR^1).

Aldehydes are abbreviated as RCHO not RCOH.

Ketones are abbreviated as RCOR' not ROR'. The C bearing the O cannot have any H's attached., (i.e., RCH$_2$OR' represents an ether, not a ketone. RCH$_2$COR' represents a ketone.)

Do not assume that the longest straight chain is written in a straight line. For example, the longest straight chain in the following compound is shown in bold and is a 9-carbon chain.

$$CH_3CH_2 - \mathbf{CH} - \mathbf{CH_2CH_2CH_2CH_2CH_3}$$
$$| $$
$$\mathbf{CH_2CH_2CH_3}$$

SELF-EVALUATION SECTION

1. Fill in the blank space or circle the appropriate response.

Carbon, with four outer shell electrons, is able to form (1) <u>one/four</u> single (2) <u>ionic/covalent</u> bonds by sharing its electrons with other elements. The bond angles are not planar but describe a (3) <u>cubic/tetrahedronal</u> shape. The remarkable ability of carbon to bond to (4) <u>itself/only metals</u> leads to the possibility of long chain compounds, ring compounds, and structures containing single, double, and even triple bonds. A single bond consists of (5) _____ electrons, a double bond of (6) _____ electrons, and a triple bond of (7) _____ electrons. For example, in the formula C$_2$H$_2$, each carbon atom is joined to two hydrogen atoms by a (8) _____ covalent bond, and the two carbon atoms are joined together by a (9) _____ bond.

2. Draw the Lewis structure of carbon tetrabromide, acetylene, and methane. Keep in mind the possibility of multiple bonds.

    Do your structures here.

3. Draw the Lewis structures for 2-iodobutane and pentane.

    Do your structures here.

4. Match the structural formulas and the names for the isomers of hexane. Look for the longest continuous chain of carbon atoms and then the location of the alkyl groups along the chain.

Formulas	Names
(1) $CH_3-CH_2-CH_2-CH_2-CH_2-CH_3$	3-methylpentane
(2) $CH_3-CH_2-CH_2-\underset{\underset{CH_3}{\vert}}{CH}-CH_3$	n-hexane
(3) $CH_3-CH_2-\underset{\underset{CH_3}{\vert}}{CH}-CH_2-CH_3$	2,3-dimethylbutane
(4) $CH_3-\underset{\underset{CH_3}{\vert}}{CH}-\underset{\underset{CH_3}{\vert}}{CH}-CH_3$	2-methylpentane
(5) $CH_3-\underset{\underset{CH_3}{\vert}}{\overset{\overset{CH_3}{\vert}}{CH}}-CH_2-CH_3$	2,2-dimethylbutane

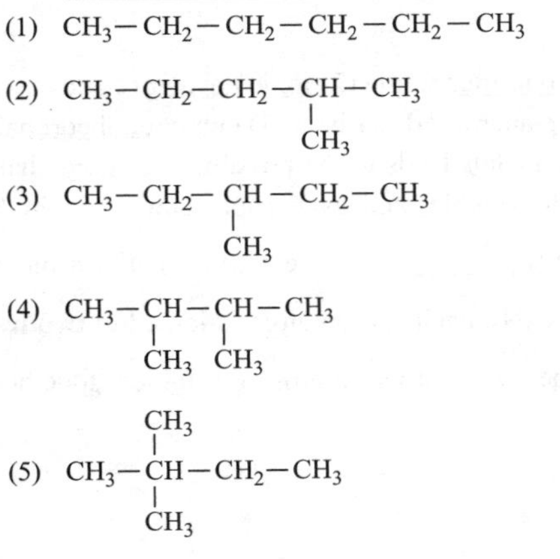

5. Name the following alkanes and alkyl halides. Look for the longest carbon chain and name them so that the substituent numbering is as small as possible.

(1) 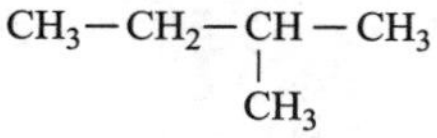 $CH_3-CH_2-CH-CH_3$
              |
             CH_3

_____

(2) $CH_3-CH_2-\overset{\overset{\displaystyle CH_3}{|}}{\underset{\underset{\displaystyle CH_3}{|}}{C}}-\overset{\overset{\displaystyle CH_3}{|}}{CH}-CH_3$

_____

(3) $CH_3-CH_2-\overset{\overset{\displaystyle Cl}{|}}{CH}-CH_3$

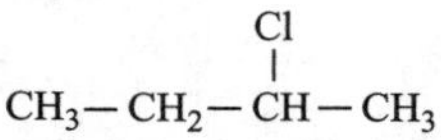

_____

(4) 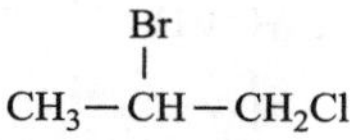 $CH_3-\overset{\overset{\displaystyle Br}{|}}{CH}-CH_2Cl$

_____

(5) 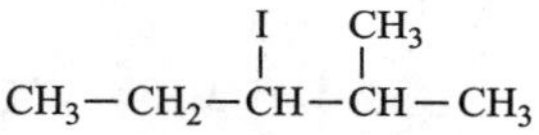 $CH_3-CH_2-\overset{\overset{\displaystyle I}{|}}{CH}-\overset{\overset{\displaystyle CH_3}{|}}{CH}-CH_3$

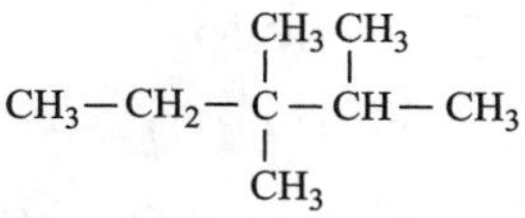

_____

(6) 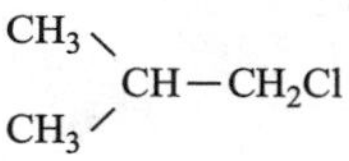 $\begin{matrix} CH_3 \diagdown \\ \quad\quad CH-CH_2Cl \\ CH_3 \diagup \end{matrix}$

_____

(7) 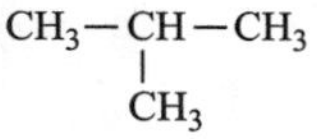 $CH_3-CH-CH_3$
              |
             CH_3

_____

(8) 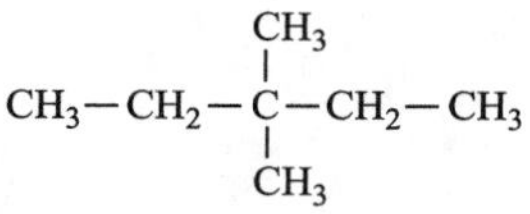 $CH_3-CH_2-\overset{\overset{\displaystyle CH_3}{|}}{\underset{\underset{\displaystyle CH_3}{|}}{C}}-CH_2-CH_3$

_____

6. Draw the possible structures for the monohydroxy (-OH) isomers of n-heptane. Be sure to eliminate any duplicate structures.

Write your structures here.

7. Fill in the missing members of the alkane, alkyne and aldehyde homologous series. Give the formulas or names or both. Note that there are two formula columns for the aldehydes. Aldehydes are more commonly written as RCHO where the R is an alkyl group. Note the trend of how the "$C_{\#}H_{\#}$" before the CHO changes.

Alkanes		Alkynes		Aldehydes		
methane	CH_4	Must have $\geq$ 2 C's		_____	HCHO	CH_2O
ethane	C_2H_6	ethyne	C_2H_2	_____	CH_3CHO	C_2H_4O
_____	C_3H_8	propyne	C_3H_4	propanal	C_2H_5CHO	C_3H_6O
butane	_____	_____	_____	butanal	_____	_____
pentane	_____	_____	C_5H_8	pentanal	_____	_____
_____	C_6H_{14}	hexyne	_____	_____	$C_5H_{11}CHO$	_____
_____	_____	_____	C_7H_{12}	_____	$C_6H_{13}CHO$	_____
octane	C_8H_{18}	_____	_____	octanal	_____	_____
nonane	_____	nonyne	_____	_____	$C_8H_{17}CHO$	_____
_____	$C_{10}H_{22}$	_____	_____	_____	_____	$C_{10}H_{20}O$

8. Name the following compounds.

(1) $CH_3-CH=CH_2$ _____

(2) $CH_3-CH_2-CH_2-C\equiv CH$ _____

(3)
$$\begin{array}{c} \quad\;\; CH_3\;\; CH_3 \\ \quad\;\;\; | \qquad | \\ CH_2=C-CH-CH_3 \end{array}$$
_____

(4)
$$\begin{array}{c} \quad\;\; CH_3 \\ \quad\;\;\; | \\ CH_3-CH-CH_2-CH_2-CH=CH_2 \end{array}$$
_____

(5)
$$\begin{array}{c} CH_3-CH_2-CH=C-CH_3 \\ \qquad\qquad\qquad | \\ \qquad\qquad\quad CH_3 \end{array}$$
_____

(6) $CH_3-CH_2-CH=CH_2$ _____

9. Name the aromatic compounds, using an appropriate system. You might choose to use either the numbering system or the o, m, p system if there are two substitutents on the benzene ring.

(1)

_____

(2)

_____

(3)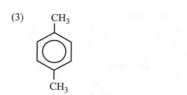

(4)

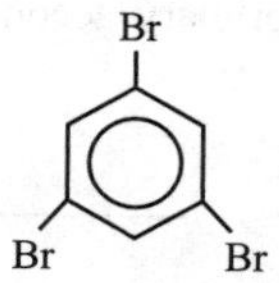

_____

_____

(5)

(6)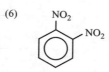

_____

_____

10. Draw the structures for the following compounds.

   (1) bromobenzene   _____

   (2) ethylbenzene   _____

   (3) phenol   _____

   (4) *p*-bromoaniline   _____

   (5) o-chloronitrobenzene   _____

   (6) 1,3,5-trinitrobenzene   _____

11. Draw the structural formulas for the following alcohols.

   (1) 2-methyl-2-butanol

   (2) 3-chloro-l-butanol

   (3) 2,2-dimethyl-l-propanol

   (4) 3-methyl-2-butanol

   (5) 2,3-dimethyl-l-pentanol

12. Identify the following alcohols as primary, secondary, tertiary, or polyhydroxy.

(1) 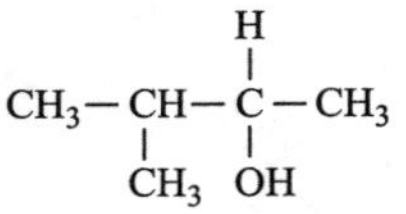
$$CH_3-\underset{\underset{CH_3}{|}}{CH}-\overset{\overset{H}{|}}{\underset{\underset{OH}{|}}{C}}-CH_3$$

_____

(2) 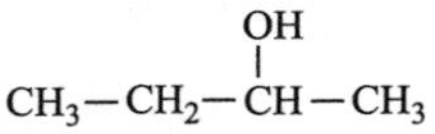
$$CH_3-CH_2-\overset{\overset{OH}{|}}{CH}-CH_3$$

_____

(3) $HO-CH_2-CH_2-OH$

_____

(4) $CH_3CH_2CH_2CH_2OH$

_____

(5) 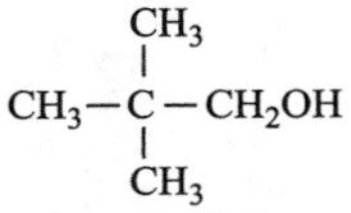
$$CH_3-\overset{\overset{CH_3}{|}}{\underset{\underset{CH_3}{|}}{C}}-CH_2OH$$

_____

(6) 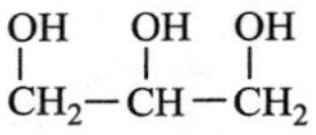
$$\overset{\overset{OH}{|}}{CH_2}-\overset{\overset{OH}{|}}{CH}-\overset{\overset{OH}{|}}{CH_2}$$

_____

(7) CH_3-OH

_____

(8) 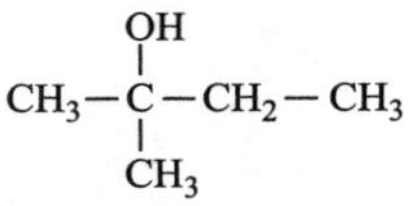
$$CH_3-\overset{\overset{OH}{|}}{\underset{\underset{CH_3}{|}}{C}}-CH_2-CH_3$$

_____

(9) 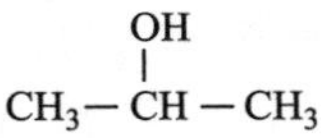
$$CH_3-\overset{\overset{OH}{|}}{CH}-CH_3$$

_____

(10) 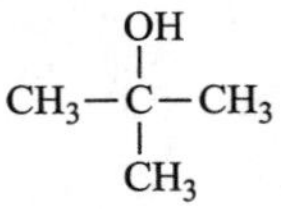
$$CH_3-\overset{\overset{OH}{|}}{\underset{\underset{CH_3}{|}}{C}}-CH_3$$

_____

13. Name the first five alcohols of question 12.

Write your answers below.

14. Draw all possible isomers of the alcohol with the formula C_4H_9OH and name them.

    Write your answers below.

15. Fill in the blank space or circle the appropriate response.

    Ethanol is one of our oldest and best known (1) <u>ketones/alcohols</u>. The preparation by the process known as (2) _____ is still carried out today. The raw materials are usually (3) _____ and (4) _____ . A biological catalyst, called a(n) (5) <u>soap/enzyme</u>, is employed in natural fermentation to convert the raw materials into ethanol and carbon dioxide. As a drug, ethanol has been shown to be a (6) <u>depressant/stimulant</u>, which is contrary to many people's beliefs. For nonfood uses, ethanol is made unfit for drinking by a process called (7) _____ . This process in effect poisons the ethanol.

16. Give the names for the following compounds.

    (1) $CH_3 - CH_2 - O - CH_2 - CH_3$

    (2) $CH_3 - O - CH_2 - CH_3$

    (3) $CH_3 - CH_2 - O - CH_2 - CH_3$
    $$\underset{\underset{CH_3}{|}}{}$$

17. Match the name of the functional group with the generalized formula.

    (1) alkane _____   a. RX

    (2) aldehyde _____   b. $R - \overset{\overset{\textstyle O}{\|}}{C} - R$

    (3) alkyl halide _____   c. $R - H$

    (4) ether _____   d. $R - CH = CH_2$

    (5) ketone _____   e. $R - OH$

    (6) alcohol _____   f. $R - C \equiv CH$

    (7) alkyne _____   g. $R - O - R$

    (8) alkene _____   h. $R - CHO$

18. After looking at the following list of structural formulas, match the names with the various formulas. Some of the names will be IUPAC and some will be common names.

(1) CH_3-O-CH_3 _____   a.   ethanol

(2) $H_2C=O$ _____   b.   formaldehyde

(3) CH_3CH_2OH _____

$$\begin{array}{c} O \\ \| \end{array}$$

(4) 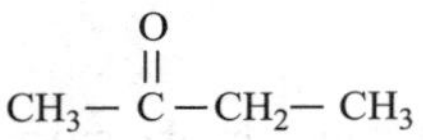 $CH_3-\overset{\overset{\displaystyle O}{\|}}{C}-CH_2-CH_3$ _____   c.   t-butyl alcohol

(5) 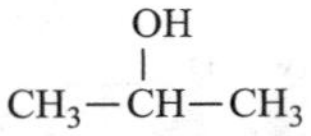 $CH_3-\overset{\overset{\displaystyle OH}{|}}{C}H-CH_3$ _____   d.   methoyxmethane

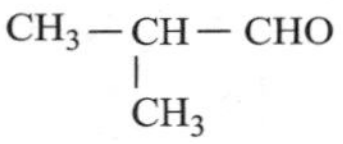

(6) $HOCH_2CH_2OH$ _____   e.   1,2-ethanediol

(7) $CH_3(CH_2)_3CH_2OH$ _____   f.   3-methyl-2-butanol

(8) 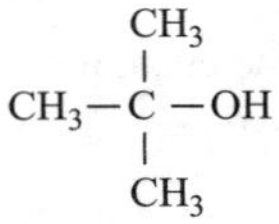 $CH_3-\overset{\overset{\displaystyle CH_3}{|}}{\underset{\underset{\displaystyle CH_3}{|}}{C}}-OH$ _____   g.   butanal

(9) 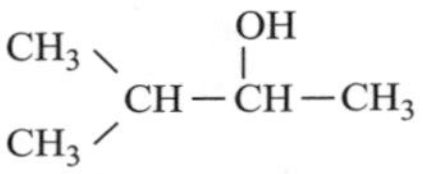 $\begin{array}{c} CH_3 \\ \diagdown \\ \\ CH_3 \diagup \end{array} CH-\overset{\overset{\displaystyle OH}{|}}{C}H-CH_3$ _____   h.   2-methylpropanal

(10) $CH_3(CH_2)_2CHO$ _____   i.   1-pentanol

(11) $CH_3-\underset{\underset{\displaystyle CH_3}{|}}{C}H-CHO$ j.   methyl ethyl ketone

_____   k.   2-propanol

19. From the list of formulas given for question 18, identify each one as an alcohol, ether, aldehyde, or ketone.

(1) _____   (2) _____

(3) _____   (4) _____

(5) _____   (6) _____

(7) _____   (8) _____

(9) _____   (10) _____

(11) _____

20. Name the organic acids using the IUPAC system.

(1) 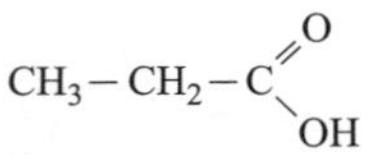 $CH_3-CH_2-C\overset{\displaystyle \diagup O}{\diagdown OH}$ (2) $CH_3-C\overset{\displaystyle \diagup O}{\diagdown OH}$

_____   _____

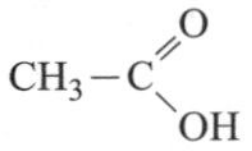

(3)

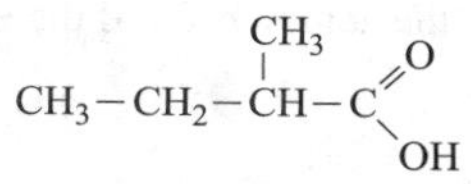

$$CH_3-CH_2-\underset{\underset{\displaystyle }{|}}{\overset{\displaystyle CH_3}{CH}}-C\overset{\displaystyle O}{\underset{\displaystyle OH}{}}$$

(4)

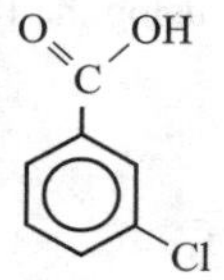

$$O=\overset{OH}{\underset{}{C}}$$ (benzene ring with Cl)

_____   _____

(5)

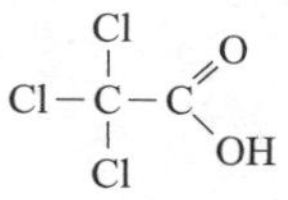

$$Cl-\underset{\underset{\displaystyle Cl}{|}}{\overset{\overset{\displaystyle Cl}{|}}{C}}-C\overset{\displaystyle O}{\underset{\displaystyle OH}{}}$$

(6)

$$O=\overset{OH}{\underset{}{C}}$$ (benzene ring with OH)

_____   _____

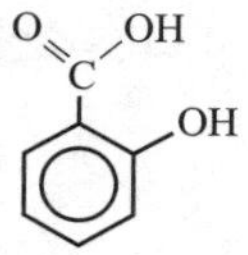

21. Fill in the blank spaces with the chemical formula or the name of the following organic acids.

	Name	**Formula**
(1)	formic acid	_____
(2)	_____	CH_3COOH
(3)	_____	CH_3CH_2COOH
(4)	salicylic acid	_____
(5)	benzoic acid	_____

22. Write the structural formulas for:

(1) Ethyl ethanoate

(2) Propyl methanoate

(3) Methyl benzoate

(4) Isopropyl ethanoate

23. Name the following esters, using the IUPAC system. Remember to name the acid portion of the ester last.

(1)

$$CH_3-C\overset{O}{\underset{O-(CH_2)_4CH_3}{<}}$$

_____

(2)

$$CH_3-(CH_2)_2C\overset{O}{\underset{O-CH_2CH_3}{<}}$$

_____

(3)

$$HC\overset{O}{\underset{O-CH_2CH\overset{CH_3}{\underset{CH_3}{|}}}{<}}$$

_____

24. Match the monomer molecules listed in the left-hand column with the generalized formulas for the corresponding polymers in the right-hand column. There may be more than one monomer associated with a polymer, as in a copolymer.

(1) $CH_2 = CCl_2$ _____

(2) $CH_2 = CH$ _____
    $\quad\quad\quad |$
    $\quad\quad\quad CN$

(3)
    $\quad\quad CH_3 \quad O$
    $\quad\quad |$
    $CH_2 = C - C$ _____
    $\quad\quad\quad\quad\quad OCH_3$

(4) $CH_2 = CH$

    _____

(5) $CH_2 = CF_2$ _____

(6)
    $\quad\quad\quad\quad H$
    $CH_2 = C$ _____
    $\quad\quad\quad\quad CH_3$

a. $\text{+}CF_2-CF_2\text{+}_n$

b. $\text{+}CH_2-CH\text{+}_n$

c. $\text{+}CH_2-\overset{Cl}{\underset{Cl}{C}}\text{+}_n$

d.
    $\quad\quad\quad\quad CH_3$
    $\quad\quad\quad\quad |$
    $\text{+}CH_2-C\longrightarrow_n$
    $\quad\quad\quad\quad |$
    $\quad\quad\quad\quad COOCH_3$

e.
    $\quad\quad\quad\quad H$
    $\quad\quad\quad\quad |$
    $\text{+}CH_2-C\longrightarrow n$
    $\quad\quad\quad\quad |$
    $\quad\quad\quad\quad CH_3$

f.
    $\text{+}OCH_2CH_2O-\overset{O}{\overset{||}{C}}$$\overset{O}{\overset{||}{C}}\text{+}_n$

g. $\text{+}CH_2-CH\text{+}_n$
    $\quad\quad\quad\quad |$
    $\quad\quad\quad\quad CN$

h. $\text{+}CH_2-\overset{CH_3}{\underset{CH_3}{C}}\text{+}_n$

— 206 —

Challenge Problem

25. Which of the following names are incorrect? (Hint, you may want to draw the compound for those that are not already shown.)

(1) 4-n-butyl-6-ethyl-5,5-dimethyl-4-octanol

(2)

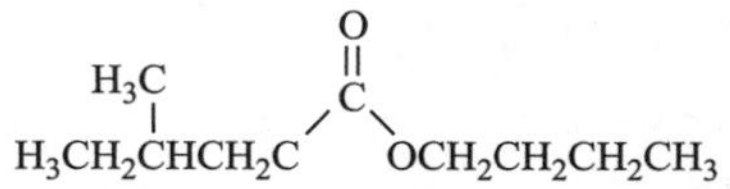

$$H_3CH_2CHCH_2C \overset{\overset{\displaystyle H_3C}{|}}{\underset{\overset{\displaystyle C}{\overset{\|}{O}}}{}} \quad OCH_2CH_2CH_2CH_3$$

2-methylbutyl butanoate

(3)

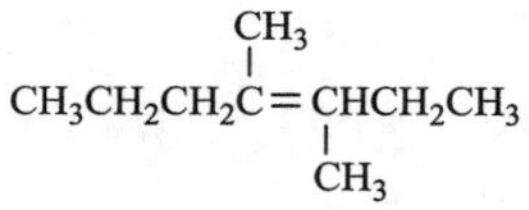

$$CH_3CH_2CH_2C\overset{\overset{\displaystyle CH_3}{|}}{=}CHCH_2CH_3 \atop \underset{\displaystyle CH_3}{|}$$

3, 4-dimethyl-3-heptane

RECAP SECTION

Chapter 19 has given you a brief look at the chemistry of carbon. Since most of the synthetic chemicals that influence our lives directly are organic chemicals, a short chapter hardly does justice to the whole field. However, you have seen that in spite of the tremendous variety of organic chemicals, chemists have managed to organize the subject and to systematize the nomenclature of organic compounds. Today, organic chemists study reaction mechanisms, molecular structure, and intricate synthesis schemes. An introduction to common functional groups and their names is important to you. As you continue in science, you will find yourselves regularly confronted with organic chemicals.

ANSWERS TO QUESTIONS AND PROBLEMS

1. (1) 4   (2) covalent   (3) tetrahedral   (4) itself
   (5) 2   (6) 4   (4) 6   (8) single
   (9) triple

2. Carbon tetrabromide

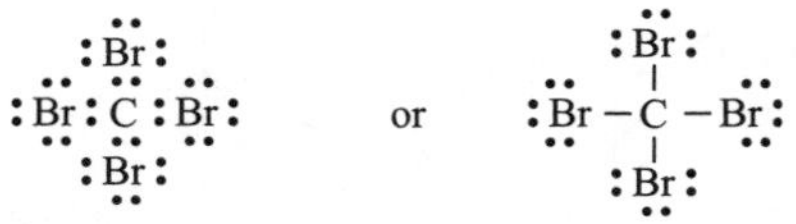

Acetylene

$$H : C ::: C : H \quad \text{or} \quad H - C \equiv C - H$$

Methane

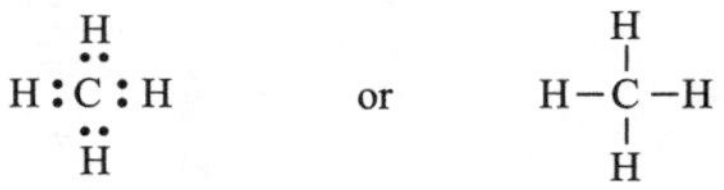

3. 2-iodubutane

```
        H  :Ï:  H  H                                H  :Ï:  H  H
        ··  ··  ··  ··                              |   |   |  |
   H : C : C : C : C : H        or           H — C — C — C — C — H
        ··  ··  ··  ··                              |   |   |  |
        H   H   H   H                              H   H   H  H
```

pentane

```
        H  H    H    H  H                          H  H  H  H  H
        ··  ··  ··   ··  ··                         |  |  |  |  |
   H : C : C  : C :  C : C : H      or        H — C — C — C — C — C — H
        ··  ··  ··   ··  ··                         |  |  |  |  |
        H  H    H    H  H                          H  H  H  H  H
```

4. (1) *n*-hexane   (2) 2-methylpentane   (3) 3-methylpentane
    (4) 2,3-dimethylbutane   (5) 2,2-dimethylbutane

5. (1) 2-methylbutane   (2) 2,3,3-trimethylpentane
    (3) 2-chlorobutane   (4) 2-bromo-1-chloropropane
    (5) 3-iodo-2-methylpentane   (6) 1-chloro-2-methylpropane
    (7) 2-methylpropane   (8) 3,3-dimethylpentane

6. Four possible isomers of heptanol (where the carbon chain is unbranched).
    There are seven carbons on heptane so all the possible structures would be as follows:

$$
\begin{array}{c}
\quad\quad\quad 6 \quad\; 5 \quad\; 4 \quad\; 3 \quad\; 2 \\
\quad\quad (OH)(OH)(OH)(OH)(OH) \\
\;\;7\quad\;\; |\quad\; |\quad\; |\quad\; |\quad\; |\quad\quad\;\; 1 \\
(HO) - C - C - C - C - C - C - C - (OH) \\
\quad\quad |\quad\; |\quad\; |\quad\; |\quad\; |\quad\; |\quad\; |
\end{array}
$$

The parentheses around the symbol for OH, are used to indicate all the possible locations to place a hydroxyl group on the heptane chain. There are seven structures to look at for duplicates. Numbers 1 and 7 are identical since the molecule is the same when looked at from either end. Numbers 2 and 6 are identical, as are 3 and 5. Number 4 is unique. Therefore there are four possible isomers of the monohydroxide of heptane (heptanol).

$CH_3\text{-}CH_2\text{-}CH_2\text{-}CH_2\text{-}CH_2\text{-}CH_2\text{-}CH_2OH$
$CH_3\text{-}CH_2\text{-}CH_2\text{-}CH_2\text{-}CH_2\text{-}CHOH\text{-}CH_3$
$CH_3\text{-}CH_2\text{-}CH_2\text{-}CH_2\text{-}CHOH\text{-}CH_2\text{-}CH_3$
$CH_3\text{-}CH_2\text{-}CH_2\text{-}CHOH\text{-}CH_2\text{-}CH_2\text{-}CH_3$

7. Members that were already in the table are shown in italics.

Alkanes		**Alkynes**		**Aldehydes**		
methane	CH_4	*Must have ≥ 2 C's*		methanal*	*HCHO*	CH_2O
ethane	C_2H_6	*ethyne*	C_2H_2	ethanal**	CH_3CHO	C_2H_4O
propane	C_3H_8	*propyne*	C_3H_4	*propanal*	C_2H_5CHO	C_3H_6O
butane	C_4H_{10}	butyne	C_4H_6	*butanal*	C_3H_7CHO	C_4H_8O
pentane	C_5H_{12}	pentyne	C_5H_8	*pentanal*	C_4H_9CHO	$C_5H_{10}O$
hexane	C_6H_{14}	*hexyne*	C_6H_{10}	hexanal	$C_5H_{11}CHO$	$C_6H_{12}O$
heptane	C_7H_{16}	heptyne	C_7H_{12}	heptanal	$C_6H_{13}CHO$	$C_7H_{14}O$
octane	C_8H_{18}	octyne	C_8H_{14}	*octanal*	$C_7H_{15}CHO$	$C_8H_{16}O$
nonane	C_9H_{20}	*nonyne*	C_9H_{16}	nonanal	$C_8H_{17}CHO$	$C_9H_{18}O$
decane	$C_{10}H_{22}$	decyne	$C_{10}H_{18}$	decanal	$C_9H_{19}CHO$	$C_{10}H_{20}O$

* methanal is more commonly known as formaldehyde
** ethanal is more commonly known as acetaldehyde
Notice the trend in each of the series: each successive member has a "CH_2" added to it regardless of series. E.g., "meth" to "eth" differ by one C and two H's, whether is it methane to ethane, or methanal to ethanal.

8.   (1)   propene   (2)   1-pentyne   (3)   2,3-dimethyl-1-butene
      (4)   5-methyl-1-hexene   (5)   2-methyl-2-pentene   (6)   1-butene

9.   (1)   ethylbenzene   (2)   ortho-dichlorobenzene or 1,2-dichlorobenzene
      (3)   para-dimethylbenzene or 1,4-dimethylbenzene   (4)   1,3,5-tribromobenzene
      (5)   toluene or methylbenzene   (6)   ortho-dinitrobenzene or 1,2-dinitrobenzene

10.   (1)   Br   (2) CH_2CH_3 (3)   OH

       (4) NH_2 (5)   (6) NO_2

11. (1)

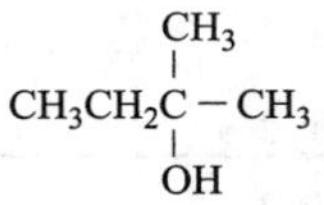

$$CH_3CH_2\overset{\overset{\displaystyle CH_3}{|}}{\underset{\underset{\displaystyle OH}{|}}{C}} - CH_3$$

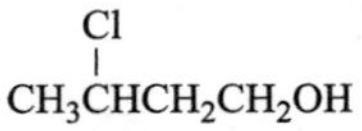

(2)

$$CH_3\overset{\overset{\displaystyle Cl}{|}}{CH}CH_2CH_2OH$$

(3)

$$CH_3 - \overset{\overset{\displaystyle CH_3}{|}}{\underset{\underset{\displaystyle CH_3}{|}}{C}}CH_2OH$$

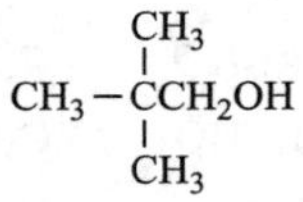

(4)

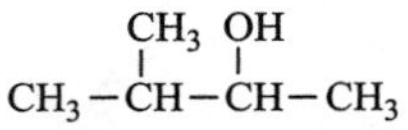

$$CH_3 - \overset{\overset{\displaystyle CH_3}{|}}{CH} - \overset{\overset{\displaystyle OH}{|}}{CH} - CH_3$$

(5)

$$CH_3 - CH_2\overset{\overset{\displaystyle CH_3}{|}}{CH} - \overset{\overset{\displaystyle CH_3}{|}}{CH} - CH_2CH$$

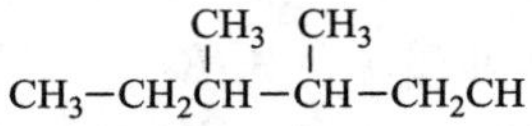

12. (1) secondary   (2) secondary   (3) polyhydroxy
    (4) primary   (5) primary   (6) polyhydroxy
    (7) primary   (8) tertiary   (9) secondary
    (10) tertiary

13. (1) 3-methyl-2-butanol   (2) 2-butanol (3) 1,2-ethanediol
    (4) n-butanol   (5) 2,2-dimethylpropanol

14. There are four possible isomers with a formula of C_4H_9OH:

(1) $CH_3CH_2CH_2CH_2OH$
    1-butanol

(2) 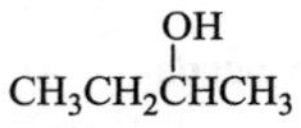
$$CH_3CH_2\overset{\overset{\displaystyle OH}{|}}{CH}CH_3$$
    2-butanol

(3)
$$CH_3 - \overset{\overset{\displaystyle OH}{|}}{\underset{\underset{\displaystyle CH_3}{|}}{C}} - CH_3$$ 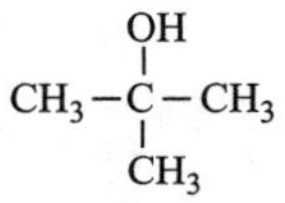
    2-methyl-2-propanol

(4) 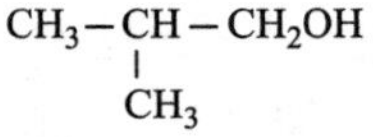
$$CH_3 - \overset{\overset{\displaystyle}{}}{CH} - CH_2OH \atop \underset{\displaystyle CH_3}{|}$$
    2-methyl-1-propanol

15. (1) alcohols   (2) fermentation   (3) starch
    (4) sugar   (5) enzyme   (6) depressant
    (7) denaturing

16. (1) diethyl ether
    (2) methyl ethyl ether
    (3) ethyl isopropyl ether

17. (1) c   (2) h   (3) a   (4) g
    (5) b   (6) e   (7) f   (8) d

18.
(1)	d	(7)	i
(2)	b	(8)	c
(3)	a	(9)	f
(4)	j	(10)	g
(5)	k	(11)	h
(6)	e		

19.
(1)	ether	(7)	alcohol
(2)	aldehyde	(8)	alcohol
(3)	alcohol	(9)	alcohol
(4)	ketone	(10)	aldehyde
(5)	alcohol	(11)	aldehyde
(6)	alcohol (diol)		

20.
(1) propanoic acid   (2) ethanoic acid or acetic acid
(3) 2-methylbutanoic acid   (4) 3-chlorobenzoic acid
(5) trichloroethanoic acid   (6) 2-hydroxybenzoic acid

21.
(1) HCOOH   (2) acetic acid

(3) propionic acid   (4)

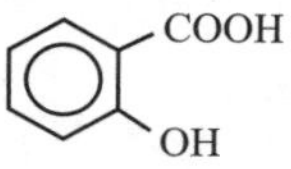

(5)

22.
(1) 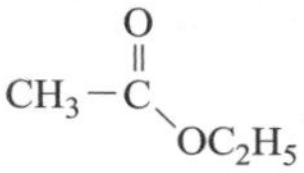 $CH_3-\overset{\overset{O}{\|}}{C}\diagdown_{OC_2H_5}$ (2) 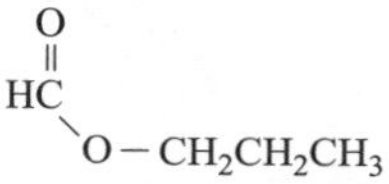 $HC\overset{\overset{O}{\|}}{}\diagdown_{O-CH_2CH_2CH_3}$ (3)

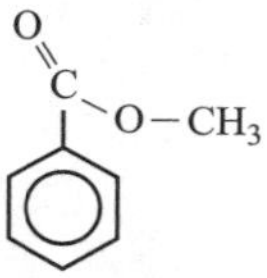

(4) 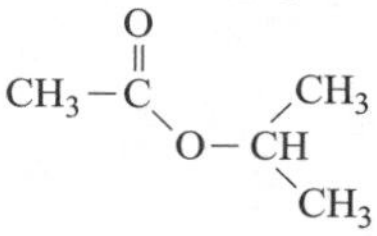 $CH_3-\overset{\overset{O}{\|}}{C}\diagdown_{O-CH}\diagup^{CH_3}\diagdown_{CH_3}$

23.
(1) pentyl ethanoate   (2) ethyl butanoate
(3) isobutyl methanoate

24.
(1)	c	(2)	g	(3)	d
(4)	b	(5)	a	(6)	e

25.
(1)

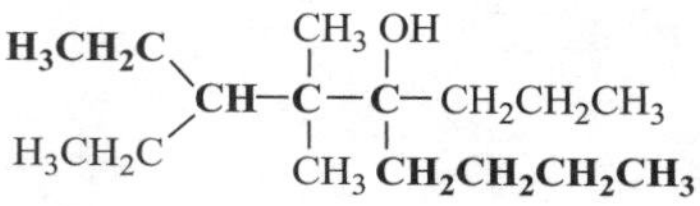

3-ethyl-4,4-dimethyl-5-n-propyl-5-nonanol
(the correct main chain is in bold)

(2)

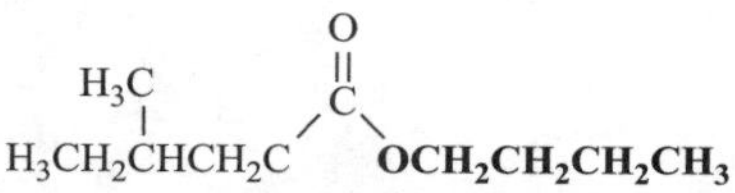

butyl 4-methylpentanoate
(the alcohol portion is in bold and is named first)

(3) This name is correct.

— 211 —

CHAPTER TWENTY

Introduction to Biochemistry

SELECTED CONCEPTS REVISITED

A very brief summary of some biochemical compounds covered in this chapter are:

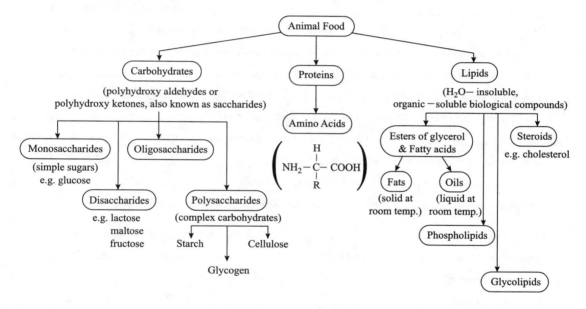

COMMON PITFALLS TO AVOID

Peptide chains are named from N-terminal to C-terminal. Be sure to look for the free NH_2 or NH_3^+ group at the end of the chain. Remember that the sequence to look for to verify that you are following the peptide chain (actual atoms to follow are in bold) is - "**N**" - "α**C**—**H**" - "**C**=**O**" - "**N**" - "α**C**—**H**" - "**C**=**O**"- (nitrogen, alpha carbon, carbonyl). If you find something else, you have ventured down a side chain.

1. Fill in the blank space or circle the appropriate response.

    The branch of chemistry that deals with the chemical reactions of living organisms and is called (1) _____ is one of the fastest moving and most exciting fields of science today. It wasn't very long ago that biochemists worked out the pathway by which carbon dioxide is chemically reduced by energy derived from the sun.

    The entire process of trapping the radiant energy of the sun in chemical bond energy is called (2) _____. The products of this process are (3) _____ molecules, which can be single units named (4) _____ , two single units bonded together named (5) _____ , or long chain molecules composed of many units called (6) _____ . The carbohydrates which are produced as a result of photosynthesis can be used for many purposes in an organism. They are a source of immediate energy or they may be stored in the form of starch. Green plants possess a rigid structure because of the polysaccharide, (7) _____ , which indicates that carbohydrates can play a structural role in addition to an energy source role.

    Two other important classes of food stuffs are (8) _____ and (9) _____ . The oil seeds such as peanuts, soybeans, and cottonseed are important sources of oils and fats. An important commercial product, (10) _____ , is made much the same way pioneers used to make it with lye, beef tallow, and a big iron kettle. The other energy compounds, the proteins, serve additional roles. Hair, fingernails, and connective tissue are structural proteins.

    The biological catalysts, (11) _____ , are proteins which serve a specialized vital role in metabolism. And, of course, the vegetable and animal protein we eat serves as food material. One of the important processes of digestion in the human stomach is to break down proteins into small sub-units. The single sub-units of proteins are called (12) _____ , of which 20 are commonly found in all proteins. Amino acids join together to form proteins through (13) _____ linkages, which produces molecules of great diversity. The large number of different protein molecules do not just develop by chance. It is directed by yet another group of compounds whose structure and function has recently been determined. These compounds are (14) _____, which we abbreviate as (15) _____ and (16) _____ . (17) _____ is found in the nucleus of every living cell, whereas most of the (18) _____ is found in the cell matrix, where it is involved in protein synthesis.

2. Match the name of a sugar with its structural formula, either open-chain or cyclic

(1) glucose   (4) fructose   (7) lactose
(2) maltose   (5) ribose
(3) galactose   (6) sucrose

(a)

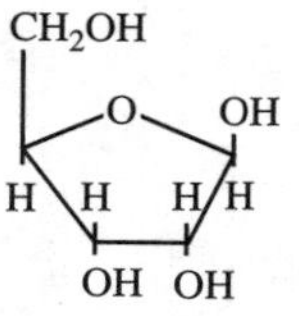

CH_2OH

(b)

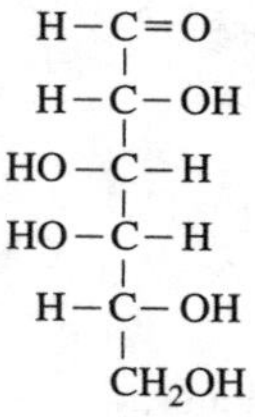

H−C=O
H−C−OH
HO−C−H
HO−C−H
H−C−OH
CH_2OH

(c)

CH_2OH

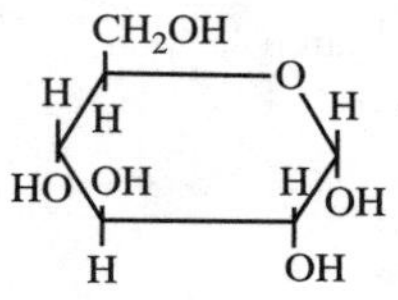

(d)

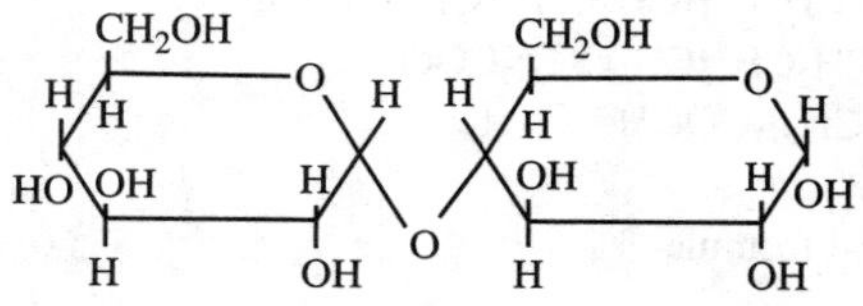

CH_2OH CH_2OH

(e)

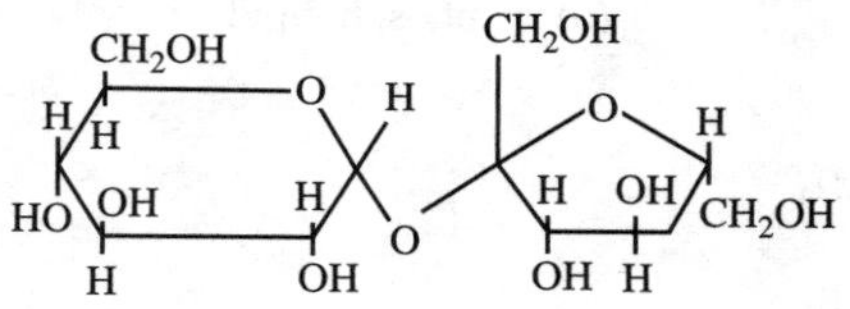

CH_2OH CH_2OH

(f)

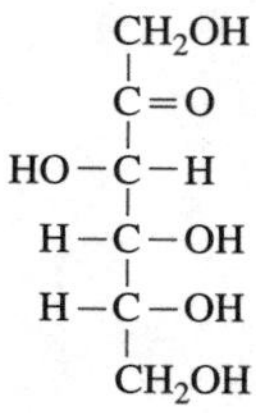

CH_2OH
C=O
HO−C−H
H−C−OH
H−C−OH
CH_2OH

(g)

CH_2OH CH_2OH

— 215 —

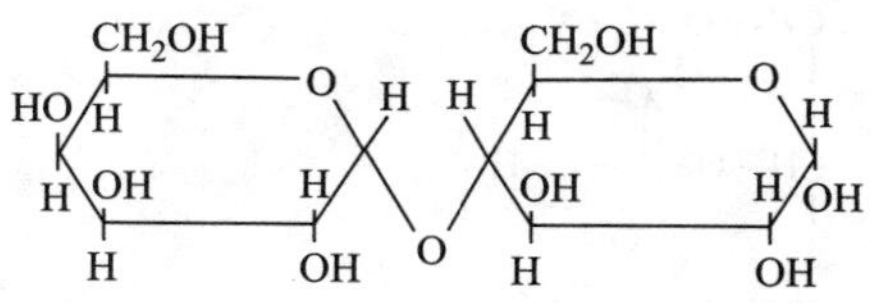

3. Classify the following carbohydrates as monosaccharides, disaccharides, or polysaccharides.

(1) galactose   (6) erythrose
(2) starch   (7) glycogen
(3) glucose   (8) maltose
(4) fructose   (9) ribose
(5) lactose   (10) sucrose

4. List the monosaccharides in the following carbohydrates:

(1) maltose   (2) lactose   (3) sucrose   (4) starch

5. Match the name of a fatty acid with its structural formula.

(1) Linoleic acid   (a) $CH_3CH_2CH=CHCH_2CH=CHCH_2CH=(CH_2)_7COOH$
(2) Oleic acid   (b) $CH_3(CH_2)_4CH=CHCH_2CH=CH(CH_2)_7COOH$
(3) Linolenic acid   (c) $CH_3(CH_2)_7CH=CH(CH_2)_7COOH$

6. Match the name of a lipid molecule with its structural formula.

(1) triacylglycerol   (2) essential fatty acid   (3) soap
(4) glycolipid   (5) steroid   (6) phospholipid

(a)

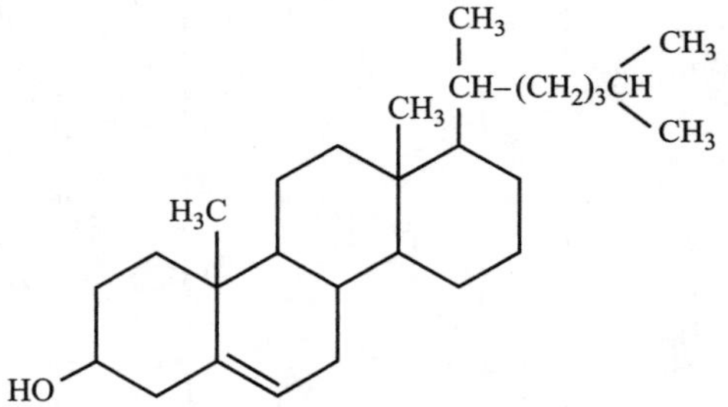

(b)

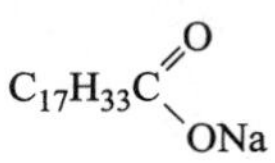

$C_{17}H_{33}C \begin{smallmatrix} O \\ ONa \end{smallmatrix}$

(c)

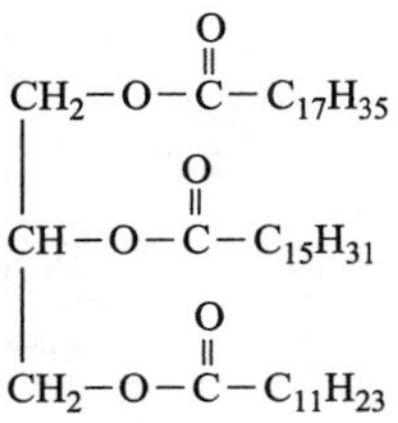

(d)

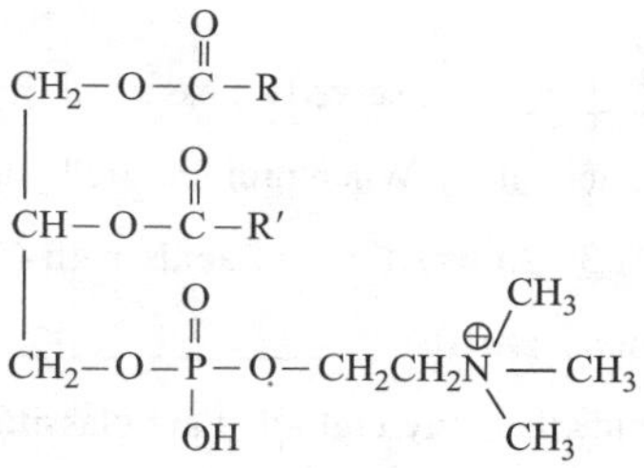

$$CH_2-O-\overset{\overset{O}{\|}}{C}-R$$

$$CH-O-\overset{\overset{O}{\|}}{C}-R'$$

$$CH_2-O-\overset{\overset{O}{\|}}{\underset{\underset{OH}{|}}{P}}-O-CH_2CH_2\overset{\overset{CH_3}{|}}{\underset{\underset{CH_3}{|}}{N}}{\overset{\oplus}{}}-CH_3$$

(e)

$$CH_3(CH_2)_{12}-CH=CH-\overset{\overset{OH}{|}}{CH}-\overset{\overset{|}{CH_2}}{CH}-NH\overset{\overset{O}{\|}}{C}-R$$

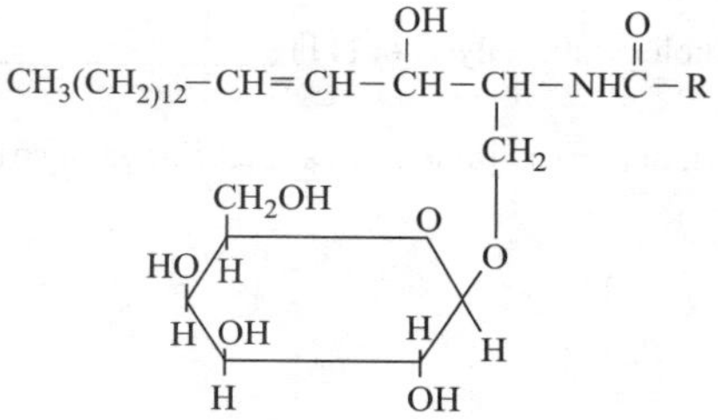

(f) $CH_3(CH_2)_4CH=CHCH_2CH=CH(CH_2)_7COOH$

7.   From the list below, circle the foods high in protein content.

bread   melon   spaghetti   apples

fish   chichen   cheese   potatoes

carrots   nuts   eggs   beans

8.   Fill in the blank space with either the name of the amino acid or its abbreviation.

Name	Abbreviation
(1) Glutamic acid	
(2)	val
(3)	ser
(4) Leucine	
(5) Cysteine	
(6) Histidine	
(7)	arg
(8)	pro
(9) Methionine	
(10) Phenylalanine	

9. Fill in the blank space or circle the appropriate response.

Proteins, which are polymers composed of (1) _____ , serve two essential functions: that

of (2) <u>a source of oxygen/a structural role</u> and (3) <u>as enzymes/as activators</u>. When proteins are hydrolyzed into

their component amino acids, scientists find approximately (4) <u>30/20</u> usual amino acids with (5) <u>two/three</u>

identical functional groups per amino acid. These two groups are (6) _____ and

(7) _____ . On a list of common amino acids there are eight that are classified as essen-

tial for (8) <u>beef/humans</u> . This means that these amino acids cannot be (9) _____ by an

individual and must be supplied in the (10) _____ . In addition to serving a variety of struc-

tural roles in an organism, proteins are also found to be biological catalysts or (11) _____ .

10. Write the two possible dipeptide structures formed when alanine is joined to valine. For each structure, point
    out the peptide linkage. Name each dipeptide.

    Write your structures here.

11. Fill in the blank space or circle the correct response.

The great variety of proteins found in any organism does not arise by chance. The substances that direct the

course of protein synthesis are known as (1) _____ . The two major types are known by

abbreviations (2) _____ and (3) _____ . There are several major

differences between these types of nucleic acids. RNA contains the sugar (4) _____ , and

DNA contains (5) _____ . DNA is a double-stranded (6) _____ ,

and RNA is only (7) _____ stranded. In addition, one of the RNA bases is different.

DNA contains the base (8) _____ and RNA contains (9) _____ . In

nucleic acids, the actual building blocks are phosphate esters. These molecules are called (10)

_____ . In the double-stranded helix of DNA, it has been found that (11) <u>complemen-</u>

<u>tary/dissimilar</u> pairing occurs between purine and pyrimidine bases. The genetic information contained in the

cell's genes is found to be the exact sequence of nucleotides along the DNA strands. When the DNA repli-

cates itself during normal cell division called (12) _____ , the two daughter cells have

the same double-stranded DNA sequences as the original cell. In human reproduction, cell splitting occurs by

a different process called (13) _____ , with the male and female both contributing to the

new cell. When fertilization takes place the zygote has a normal complement of genes, half from the mother

and half from the father. Each gene, which directs the synthesis of one protein, contains what is termed the genetic code. Each codon, for a specific amino acid, is composed of (14) _____ nucleotides in sequence. If, for some reason, the DNA structure is changed or a breakdown occurs in the protein synthesis process, a (15) _____ may occur. Most severe cases result in an organism's death, but sometimes the organism survives. For example, sickle cell anemia results from (16) <u>two/one</u> change(s) in an amino acid on one chain of hemoglobin.

12. Circle the appropriate response.

Lipids are a group of (1) <u>organic/inorganic</u> compounds classified on the basis of their solubility (i.e., they are soluble) in fat solvents such as (2) <u>ether, benzene, etc./water, salt solution, etc.</u> The three types of simple lipids are (3) _____, (4) _____, and (5) _____.

13.

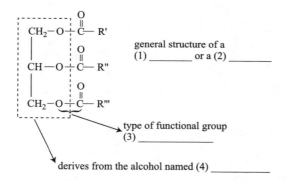

general structure of a
(1) _____ or a (2) _____

type of functional group
(3) _____

derives from the alcohol named (4) _____

14. Assign the following as relevant to a fat or an oil.

(1) solid at room temperature _____

(2) mostly from vegetable sources _____

(3) high degree of unsaturation in fatty acid groups _____

(4) mostly from animal sources _____

(5) liquid at room temperature _____

(6) often fully saturated (no double bonds in the fatty acid groups) _____

Challenge Problems

15. Name the following pentapeptides.

(1) thr-pro-ala-val-phe
(2) gly-tyr-his-leu-val

16. An octapeptide was known to contain two proline residues and two leucine residues. The following peptides were obtained after partial hydrolysis. Determine the sequence.

ala-pro-his, pro-leu, ser-leu, his-pro, pro-leu-val-ser

RECAP SECTION

Chapter 20 is a survey chapter covering the topic of biochemistry in much the same manner that Chapter 19 surveys the broad field of organic chemistry. After reading the chapter, it is important to categorize the many different compounds into as few classes as possible. Carbohydrates, lipids, and proteins can serve as energy sources or fuel molecules. Nucleic acids and enzymes serve functional roles and may act as information molecules. Proteins have the widest range of roles since they can be used strictly for fuel, employed in structural roles such as the building of hair and fingernails, or act as catalysts when enzymes.

The discussion concerning the other types of biological compounds is of general interest and would be worthwhile reading for the well-informed individual. If you are anticipating a career in an allied health field, it would be wise to do further reading and take additional courses in this area.

ANSWERS TO QUESTIONS

1. (1) biochemistry   (2) photosynthesis   (3) carbohydrate
    (4) monosaccharides   (5) disaccharides   (6) polysaccharides
    (7) cellulose   (8) lipids   (9) proteins
    (10) soap   (11) enzymes   (12) amino acids
    (13) peptide   (14) nucleic acids   (15) DNA
    (16) RNA   (17) DNA   (18) RNA

2. (1) −(c),   (2) −(d),   (3) −(b),   (4) −(f),   (5) −(a)
    (6) −(e),   (7) −(g)

3. (1) mono   (2) poly   (3) mono
    (4) mono   (5) di   (6) mono
    (7) poly   (8) di   (9) mono
    (10) di

4. (1) maltose is 2 glucose units
    (2) lactose is galactose plus glucose
    (3) sucrose is glucose plus fructose
    (4) starch is glucose units

5. (1) −(b),   (2) −(c),   (3) −(a)

6. (1) −(c),   (2) −(f),   (4) −(b),
    (4) −(e),   (5) −(a),   (6) −(d)

7. Fish, chicken, nuts, cheese, eggs, beans.

8. (1) glu   (2) Valine   (3) Serine
    (4) leu   (5) cys   (6) his
    (7) Arginine   (8) Proline   (9) met
    (10) phe

9.
(1) amino acids   (2) a structural role   (3) as enzymes
(4) 20   (5) two   (6) amino
(7) carboxyl   (8) humans   (9) synthesized
(10) diet or food   (11) enzymes

10.

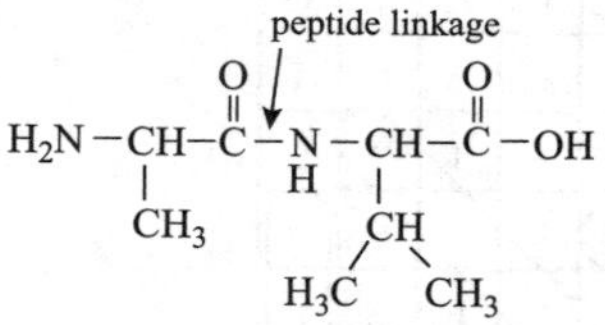

peptide linkage

$$H_2N-CH-\overset{\overset{O}{||}}{C}-\underset{\underset{CH_3}{|}}{N}-CH-\overset{\overset{O}{||}}{C}-OH$$

(alanyl structure with CH₃ on left carbon, N–H, and CH(H₃C)(CH₃) branch)

alanylvaline

peptide linkage

$$H_2N-CH-\overset{\overset{O}{||}}{C}-\overset{H}{N}-CH-\overset{\overset{O}{||}}{C}-OH$$

(valyl structure with CH(H₃C)(CH₃) on left carbon, and CH₃ branch)

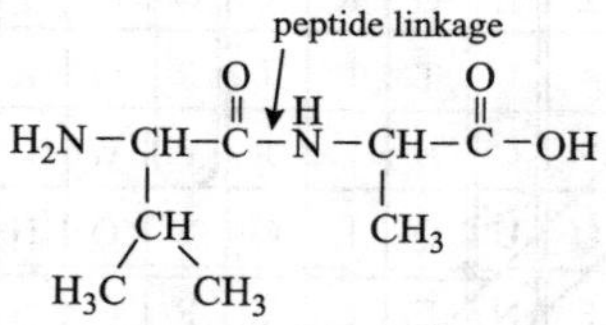

valylalanine

Note that it does make a difference which amino acid is "first." These are not identical structures!

11.
(1) nucleic acids   (11) complementary
(2) DNA   (12) mitosis
(3) RNA   (13) meiosis
(4) ribose   (14) three
(5) deoxyribose   (15) mutation
(6) helix   (16) one
(7) single
(8) thymine
(9) uracil
(10) nucleotides

12. (1) organic   (2) ether, benzene, etc.   (3), (4), (5) fats, oils, waxes

13. (1), (2) fat, oil   (3) ester   (4) glycerol

14. (1) fat   (2) oil   (3) oil   (4) fat   (5) oil   (6) fat

15. (1) threonylprolylalanylvalylphenylalanine
    (2) glycyltyrosylhistidylleucylvaline

16. The octapeptide would be: ala-pro-his-pro-leu-val-ser-leu

H	C	O	M	P	O	U	N	D	D	I	S	O	T	O	P	E	S
O	X	C	S	Z	K	L	O	Y	P	M	N	Q	R	B	L	C	T
M	P	R	U	D	K	Q	T	V	F	V	G	M	W	E	Z	T	Y
O	I	Z	E	O	R	V	O	M	T	S	E	O	C	U	A	L	G
G	U	A	L	T	E	I	R	O	R	E	S	T	T	L	A	C	I
E	I	F	C	S	U	P	P	W	L	L	R	A	B	A	I	T	U
N	G	O	U	Q	E	G	Y	O	H	O	J	Y	G	R	N	O	P
E	E	E	N	E	U	T	R	O	N	P	A	M	V	E	E	S	E
O	L	B	U	I	S	W	H	E	J	I	M	N	M	T	K	O	R
U	S	O	V	T	Z	U	G	J	F	D	V	E	C	L	D	E	I
S	I	E	I	R	K	A	O	A	B	K	L	T	U	M	L	E	O
K	N	G	S	D	T	M	T	H	V	E	P	R	S	D	D	A	D
L	L	A	T	I	B	R	O	I	P	A	D	O	I	M	E	R	I
O	E	W	V	S	L	H	F	L	O	R	G	E	U	Q	N	V	C
B	O	I	O	N	N	A	P	A	E	N	O	L	T	W	S	Z	L
M	T	V	M	E	O	S	I	D	N	C	E	M	N	O	I	N	A
Y	L	S	A	T	R	T	Z	M	A	B	U	N	A	L	T	U	W
S	P	J	D	A	T	C	I	O	T	M	J	L	E	K	Y	V	Z
A	E	O	R	B	C	M	R	P	E	B	U	M	E	R	E	I	D
B	H	T	L	C	E	A	X	T	M	X	Y	U	T	T	G	G	C
G	D	I	U	F	L	O	A	O	F	E	S	N	X	G	A	Y	V
S	T	N	E	M	E	L	E	N	O	I	T	I	S	N	A	R	T
A	M	G	W	U	P	Q	N	S	W	D	L	E	R	V	K	C	B
H	I	V	R	E	T	E	M	O	R	D	Y	H	R	W	E	J	F

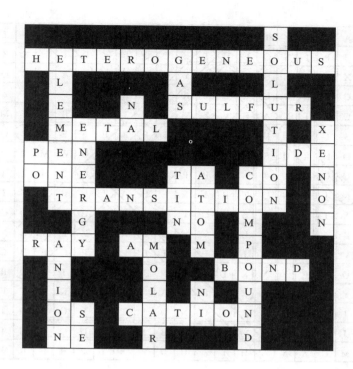

A	E	G	Y	T	F	E	H	Y	K	B	D	I	S	P	D	N	I
G	P	I	S	O	L	U	T	E	O	Q	F	Y	D	I	L	K	Z
X	T	D	N	O	A	I	S	Y	C	I	E	V	L	A	H	P	O
C	K	E	B	G	L	O	M	R	L	X	R	U	N	F	O	J	Z
U	M	T	V	A	E	V	Z	M	A	O	T	E	Q	B	Q	Y	M
S	F	A	M	H	C	R	E	D	I	E	R	E	F	F	I	B	Q
D	N	R	U	S	H	E	R	N	C	S	Q	T	U	W	M	J	P
C	O	U	I	J	A	L	B	P	T	L	C	R	C	G	H	Z	W
N	I	T	N	U	T	B	I	N	E	H	A	I	V	E	E	Y	I
X	T	A	O	W	E	I	C	T	O	L	R	T	B	A	L	O	Q
K	A	S	R	D	L	C	V	M	O	E	W	M	Z	L	V	E	N
V	R	G	D	L	I	S	E	M	T	B	L	N	E	C	E	B	L
O	T	F	Y	X	E	L	I	O	N	I	Z	A	T	I	O	N	R
Y	I	G	H	W	R	M	H	Z	A	X	U	F	A	K	T	T	F
I	T	S	V	A	X	P	U	N	O	I	T	U	L	O	S	G	J
J	P	L	Q	Z	M	U	I	R	B	I	L	I	U	Q	E	E	T
M	C	X	S	A	H	T	A	U	D	F	N	Y	P	R	B	R	W